Life is Full of Science

An Interdisciplinary and Cultural Teaching Approach

Life is Full of Science

An Interdisciplinary and Cultural Teaching Approach

First Edition

Jiyoon Yoon

University of Texas—Arlington

cognella® | ACADEMIC PUBLISHING

Bassim Hamadeh, CEO and Publisher
Jennifer McCarthy, Acquisitions Editor
Amy Smith, Project Editor
Christian Berk, Associate Production Editor
Alisa Munoz, Licensing Coordinator
Natalie Piccotti, Director of Marketing
Kassie Graves, Vice President of Editorial
Jamie Giganti, Director of Academic Publishing

ISBN: 978-1-5165-0439-8 (pbk) / 978-1-5165-0440-4 (br)

cognella® | ACADEMIC PUBLISHING

CONTENTS

PREFACE

This textbook is for preparing K–6 science teachers of diverse students. It has been reported that the teachers of young children consider science a difficult subject. This book provides opportunities to change their perceptions about science through the author's stories related to *daily life*. The stories help teachers understand science better and apply the sciences easily into their lives. Therefore, their perceptions about and attitudes toward science may be changed positively.

In this book, science is also introduced with an *interdisciplinary* approach, integrating with other subjects, like math, social studies, technology, art, and music. In real life, science comes together with other subjects. Therefore when science is presented in an interdisciplinary way, teachers enable their students to better grasp science concepts and implement their learning easily into their lives.

In order to share the skills to serve diverse student populations and build inclusive curriculum for future global classrooms, this textbook is designed by comparing and integrating the curriculum and *culture of other* countries that have high international achievements in science, like South Korea. The author has found that, to improve the science achievements of students, American science teacher education needs to enhance the quality of teacher candidates by providing diverse cultures and education systems from other countries with higher achievements in science. The author collaborated with Korean scholars to investigate and integrate science curriculum for young children in America and Korea, connecting with the science standards and culture of both countries. The methods for integrating Korean curriculum and culture with the teaching science provided in this textbook can be extended to many other cultures around the world.

Further, to enhance creativity and inquiry in science, this textbook is using the *5E learning cycle* (Engagement, Exploration, Explanation, Extension, and Evaluation). The 5E learning cycle is firmly grounded in constructivist, student-centered, inquiry-based teaching and learning, engaging students in active inquiry and problem solving activities toward understanding, applying, and extending learned concepts in the real world. The 5E learning cycle successfully supports 1) the meaning of the nature of science, because it is active and inquiry-based; 2) the purpose of school, because it promotes the development of thinking abilities; and 3) the methods of how children learn, covering their preferred learning styles: visual, auditory, kinesthetic, and tactile.

Not only does this textbook describe the integration of other cultures and curriculum and the 5E learning cycle model, but it also includes many special features:

a. U.S. National Standards/Korean teacher qualification standards
b. Hands-on, inquiry-based student activities
c. International traditional games and culture
d. Summaries, assignments, and a Tip of the Day at the end of each chapter facilitate better organization, summarization, and understanding and the application of learning

The Tip of the Day is a daily message that the author sends to the teacher candidates who are practicing teaching in schools to help them to apply their learning from each chapter.

The textbook is composed of the four key components: Educational, Assessment, Global, and Technological (see Figure 0.1):

a. The educational component supports theoretical background, instructional strategies, and issues and trends for teaching science.

b. The assessment component provides teacher candidates with the assessment tools necessary to enhance teaching science to diverse student populations.

c. The global component provides an overview of the issues, principles, and practices associated with effective teaching in culturally, linguistically, and environmentally diverse elementary science classroom contexts from the perspectives of both the teacher and the internationally diverse learners.

d. The technology component explores the use of instructional technology in science classrooms, emphasizing the utilization and evaluation of various technologies and their appropriateness for teaching science.

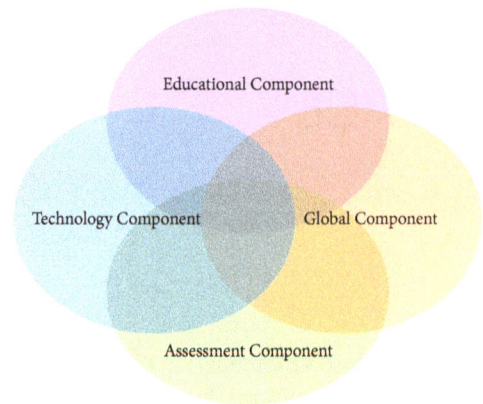

FIGURE 0-1 Key components of the textbook

1

Nature of Science

How Much Science Do We Need to Know?

How much science do we need to know as science teachers? You may have this question in mind when you start teaching. I also like to ask all of you this question first, because it gives you the opportunity to think about what science is, what has been emphasized in science education, and how many other science education organizations talk about the answer to this question.

This question reminds me of the Christmas gift my mother gave my younger sister. I already knew that my mother was Santa Claus when I was eight years old, but I did not say it to my younger sister, because I wanted to preserve her cute imagination. That Christmas night, I saw Mom place the gift near my younger sister's pillow. I was so curious about what was inside. As soon as Mom left our room, I got up and smelled the box. There was no smell. I tried to see through the box, but could not see what was inside. I shook the box, but there was no sound. I thought and thought and concluded that the gift might be clothes or a scarf, because there was no smell nor sound, and the box was light. Then, I decided not to switch my Christmas gift with my sister's, because mine was a backpack that I thought would be better for me. The next morning, I found that I was right! My sister's was a scarf that she loves! Due to the use of my five senses, I could find out what was inside and made both my sister and me happy on that Christmas morning.

Through this chapter, we will discover the definition and nature of science, characteristics of science, history of science education, and positions of the professional societies in science education to find the answer to the topic question of this chapter.

Science Activity 1-1 Mystery Bag

Duration: 20 minutes
Materials: lunch bags, staplers, a stop watch

Activity Procedure:
Ask the class to make groups (5 or 6 members in a group)

Spread out a lunch bag to each student in the class

Ask the students to go outside of the classroom, find an object, and put the object in the lunch bag *secretly*. No one but the owner of the bag can know what is inside his or her own lunch bag. If some of the students want to use their own materials in the classroom, the decision can be made by the teacher whether they will use their own materials or not.

Once the students come back to the class with their lunch bags filled with the secret objects, ask them to staple the bags one time (this is to prevent the other students from seeing what is inside).

Ask the class to exchange their bags with the *other* group members'. If they want to get their own lunch bags, ask them to write their names on their lunch bags before exchanging them.

Once they exchange their bags, a group comes out to the front of the class, lines up, and faces the class. Ask the group not to open the bags until the teacher says, "Start."

When the teacher says, "Start," the first member of the group opens his or her bag and explains what is inside without pulling the object out of the bag. Then, the rest of the group members guess what it is.

The whole duration is counted until all of the members of the group guess what are inside their bags. The winner group is the one who takes the least amount of time to guess correctly.

The Definition of Science

Science has been defined in various ways. Some people like to say science is an understanding of the world. Also, science can be defined as the information that can be measured or seen and verified by scientists. But in this textbook, the definition of science is more focused on *process*. Project 2060 (Rutherford, F. J. & Ahlgren, A., 1991) defined science as "a process for producing knowledge."

The process extends from making careful observations of phenomena to creating theories for making sense out of those observations (Rutherford, F. J. & Ahlgren, A., 1991). However, once a theory is produced, this is the start of another process.

Change in knowledge is inevitable, because new observations challenge prevailing theories. No matter how well one theory explains a set of observations, it is possible that another theory [fits] just as well or better or [fits] a still wider range of observations. In science, testing, improving, and [the] occasional discarding of theories, whether new or old, [occurs] all the time. Scientists assume that, even if there is no way to secure complete and absolute truth, increasingly accurate approximations can be made to account for the world and how it works. (Rutherford, F. J. & Ahlgren, A., 1991)

As a conclusion, science is the process for producing all this knowledge around the world.

The following Mystery Bag activity helps students understand the definition of science.

As soon as they have new bags from the other group members, the teacher may notice that many of the students are curious about what is inside the bags. Some of them might try to look through, smell, touch, and even lick the bags to find what is inside (*Curiosity*). Then, when the groups come to the front of the class, each member looks inside (*Observation*) the bag. As soon as they find the objects in the bags, they start *measurement, classification,* and *inferring of* the object and explain their findings to other group members(*Communication*). The other group members guess (*Prediction*) what is inside the bag based on their group member's explanation. After this process, the group discovers what is inside the bag (*Product of Science, "Fact" or "Theory"*). This whole process is called "science." The process that they completed through this activity is summarized as follows:

> Curiosity (at the beginning)
>
> Exploration by observation, measurement, classification, prediction, communicating, and inferring (during the activity)
>
> Product of finding what is inside the bags (at the end of the activity)

Basic Skills for Doing Science

When they guessed what was inside the bags, the students were using six basic skills:

> Observing
> Measuring
> Classifying
> Predicting
> Inferring
> Communicating

It is important for science teachers to have their students improve these six basic skills for doing science. Once the students develop these skills, they can explore and discover the world by themselves.

Observing is the skill of collecting information with the senses and describing scientific events. There are two different kinds of observations: Quantitative and Qualitative. Quantitative observation is the information that can be measureable or countable. For example, this ruler is 3 meters long, there are 4 marbles, this road is 50 kilograms, and today, it is 35 degrees Celsius. Qualitative observation is the information that can be described, but not measured. For example, I see red flowers, they smell like fresh baked cookies, and it tastes bitter.

Inferring involves developing conclusions or deductions based on observations of evidence. The mystery footprint activity is one of the activities for improving students' observation and inference skills. Figure 1-1 is the image of Mystery Footprint, and Table 1-1 is the worksheet for the Mystery Footprint activity.

Some students might confuse inferring with predicting. The predicting skill is different from the inferring skill. Through the activity above, students developed their conclusions or deductions based on the footprints they observed as evidence. The conclusions or the deductions they developed are focused on the past events or cases. The predicting skill involves guessing what will happen next based on their observations and what the students already know. The Korean folktale Heungbu and Nolbu, is an exemplary story to improve predicting skills among students.

Here is the brief summary of the tale:

Part 1

Once upon a time, there were two brothers named Heungbu and Nolbu. The younger brother, Heungbu, was good and kind, and the older brother, Nolbu, was bad and greedy. One day, their father called them up, "If I die, divide the property in half, and you each get a half."

Part 2

But after their father died, the bad Nolbu kicked Heungbu, his wife, and his nine children out without any money and food. On the road, the nine children starved to death. Heungbu went to his brother, Nolbu's house and begged for food. But the greedy Nolbu accused his brother of being lazy and refused to help him.

Part 3

Heungbu came back home empty handed. The cold winter went by and the warm spring came. A pair of swallows made their home at Heungbu's house and bred baby swallows. One day, Heungbu heard a loud cry of swallows. He saw a big snake was sticking out its tongue to eat baby swallows. Heungbu hit the snake with the stick and it went away. But a baby swallow fell when it was trying to avoid the snake and broke its leg. Heungbu put a bandage on its leg. The autumn came, and the swallow family headed South.

Science Activity 1-2 Mystery Footprint

Materials: Image of Mystery Footprint (see Figure 1-1) and Worksheet (see Table 1-1)

Activity Procedure:

At the beginning, show only Section 1 of the image, covering the two other sections, and ask the students two questions:

What do you observe? List at least three observations.

What do you infer?

After the students write down their observations and inferences, move the cover to show Section 2 (but still covering Section 3 of the image), and ask the same questions:

What do you observe? List at least three observations.

What do you infer? (What do you think now?)

After the students write down their observations and inferences, move the cover to show the whole image. And ask the same two questions again.

After the students write down their observations and inferences, ask the students to write their conclusions about what they learn from this Mystery Footprint.

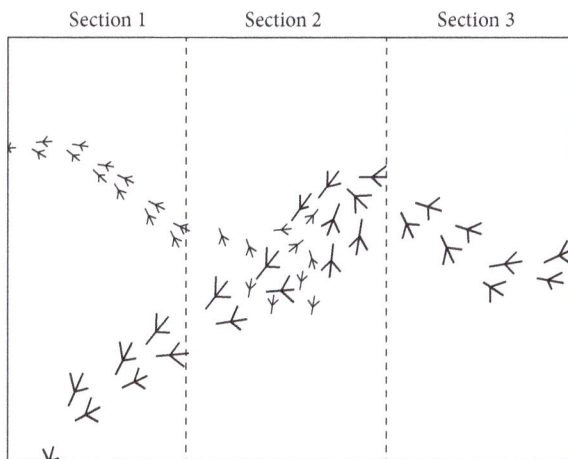

FIGURE 1-1 Mystery footprints

TABLE 1-1 Worksheet for the Mystery Footprint Activity

Mystery Footprints Activity Worksheet

Section	Observation	Inference
1		
2		
3		
Conclusion		

Part 4

Next year, spring came. Each swallow came back with a gourd seed in its mouth. Heungbu's family planted the gourd seeds and made them grow successfully. Heungbu's family sawed the gourd open and watched in astonishment. Valuable treasures poured out from the gourd. When he sawed another gourd, a house, rice, silk, and a jewel fell out.

Part 5

Nolbu heard the news and caught a swallow, broke its leg, and applied a bandage. "Bring me a lot of gourd seed," he commanded. The next year, the swallow delivered him a gourd seed. Nolbu planted it, it grew, and he quickly opened the gourd. The gourd exploded, and Nolbu's family was surprised by it. Out of it came ghosts, demons, snakes, etc., and they knocked Nolbu's family down. They took away everything Nolbu had, and he became a beggar.

Science Activity 1-3 Measuring in Kitchen

1 Making Sugar Cookies

Measure ingredients to make 48 cookies with the following recipe:

2 cups all-purpose flour
1 teaspoon baking soda
1/2 teaspoon baking powder
1 1/2 cups white sugar
1 teaspoon vanilla extract
1 cup softened butter
1 egg

Combine flour, baking soda, and baking powder. In a separate bowl, combine cream sugar, vanilla extract, and butter. Add egg and mix; then, add flour mixture.

Preheat oven to 375°F or 191°C. Bake for 10 minutes.

2. Measuring Spoons

Standard U.S. measuring spoons come in the following sizes:

- 1 tablespoon (tbsp.) = 15 milliliters (mL)
- 1 teaspoon (tsp) = 5 milliliters (mL)
- 1/2 teaspoon (tsp) = 2.5 milliliters (mL)
- 1/4 teaspoon (tsp) = 1.25 milliliters (mL)
- 1/8 teaspoon (tsp) is about .5 milliliters (mL)

In countries that use the metric system—that is, most countries other than the U.S., metric measuring spoons are the same size as U.S. measuring spoons—they are just labeled a little differently. Metric spoons are measured in milliliters, or mL, as shown above.

3. Measuring Liquid Ingredients ("Holt Science & Technology", 2000, p. 39)

In a graduated cylinder or beaker, most liquids form a meniscus, or a curved upper surface. A meniscus is caused by surface tension. When a liquid, such as water, is more attracted to the walls of the container than to itself, it curves up at the edges like a smile. When some liquids, such as mercury, are more attracted to themselves than to the walls of the container, they curve down like a frown. Read the volume of a liquid from the center of its meniscus (Figure.1-2), not from the curved edges (Figure.1-3).

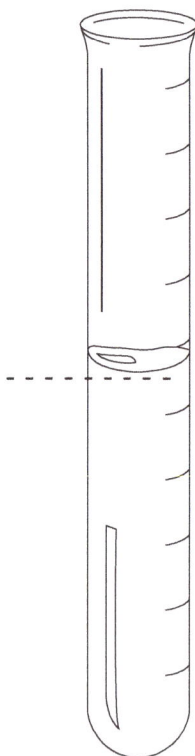

FIGURE 1-2 Correct way to read liquid volume

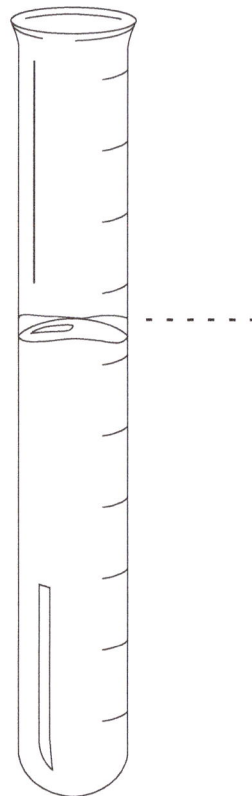

FIGURE 1-3 Not correct way to read liquid volume

4. Measuring Dry Ingredients

Dry ingredients are often measured with nested measuring cups. These include:

- 1 cup (c) = 8 ounces (oz.) = 250 milliliters (mL)
- 1/2 cup (c) = 125 milliliters (mL)
- 1/3 cup (c) is about 75 milliliters (mL)
- 1/4 cup (c) is about 50 milliliters (mL)

Part 6

When Heungbu heard the news, he asked Nolbu to live together. After that day, Nolbu became a good man and lived a blissful life.

As you read this, students can be asked to predict what will happen next. Especially, whenever the gourd cut in half, students can imagine what will come out from the gourds (in Part 4 & 5). It will be interesting to see their answers comparing Nolbu's gourd with Heungbu's.

Measuring skill involves expressing physical characteristics in quantitative ways. Measuring is "comparing an unknown quantity with a known (metric units, time, student-generated frames of reference). Observations are quantified using proper measuring devices and techniques" (American Association for the Advancement of Science, n.d.). Charts, graphs, or tables can be generated manually or with computer software.

Classifying skill is "grouping or ordering objects or events according to similarities or differences in properties. Lists, tables, or charts are generated" (American Association for the Advancement of Science, n.d.).

Communicating skill brings the rest of the basic process skills together to report to others what has been found through experimentation.

Characteristics of Science

Science is characterized by
 a. Rejecting Authority
 b. Consistency
 c. Parsimony

Science is changing all the time. Whenever science goes through the process to discover something new, previous authority is rejected by the new findings (Rejecting Authority). Also, because of this change, science is using the simplest explanation available not using details that will change soon (Parsimony). But the main ideas or main frameset will not change at all (Consistency). For example, the atom model history can explain these characteristics of science.

John Thomson proposed the plum pudding model, which represents an atom model composed of the

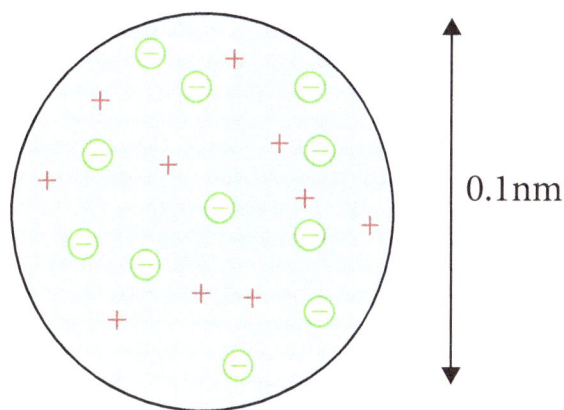

FIGURE 1-4 Thomson Atom Model

scattered positive electricity with negative particles implanted throughout the atom. For this discovery, he was awarded the Nobel Prize in physics in 1906 (see Figure 1-4).

Ernest Rutherford proposed the Rutherford's atomic model, in which the atom is composed of mostly hollow space with a positively charged nucleus (that contains protons and neutrons) concentrated at the center and surrounded by negative electrons (Figure 1-5). Rutherford received the Nobel Prize in chemistry in 1908 for his contributions to the structure of the atom. In 1913, Neils Bohr introduced that the negatively charged electrons moved along only the certain circular orbits allowed. This model of the atom helped explain the emission spectrum of the hydrogen atom. Bohr received the Nobel Prize in physics in 1922 for his theory.

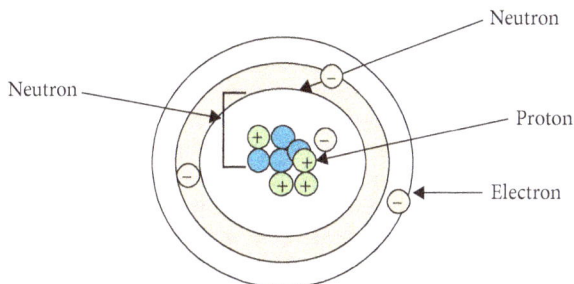

FIGURE 1-5 Solar System Model

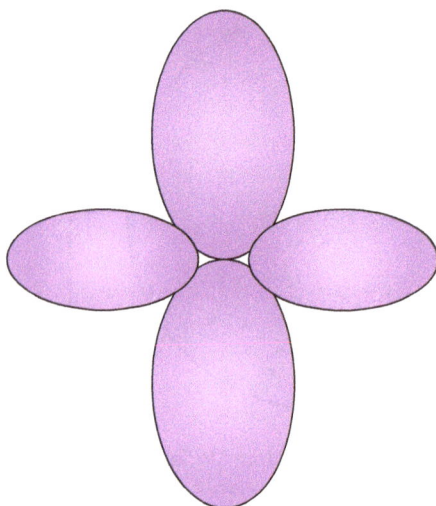

n. Clouds Model

FIGURE 1-6 Electron Clouds Model

However, in the 1920s,

Erwin Schrodinger and Werner Heisenberg developed probability functions to determine the regions, or clouds, in which electrons would most likely be found (Figure 1-6). An atom consists of a dense nucleus composed of protons and neutrons surrounded by electrons that exist in different clouds at various energy levels ("Different Atomic Models," n.d.).

All through the atom model history, the accepted details about atom models were rejected after new findings, but all the atom models consistently show that an atom is composed of positive and negative particles. So, the atom models are simple and succinct.

History of Science Education

In 1957, Russian scientists launched a satellite, Sputnik1, for the first time in the history of the world. This was a perceived failure of American science education

and engineering. The whole education system in America came under intense scrutiny. Rote learning and the tendency to reward and reinforce unoriginal thinking came under attack at the university level (Currie, 2005).

In 1959, Bruner held the Woods Hole Conference due to the threat posed by the USSR Sputnik launch. There, the conference attendees identified problem areas in science education and offered solutions. The conference marked the beginning of a new trend in science education emphasizing the *process*, rather than the product of science. The scientific community urged that students should learn science in the same way scientists do science (Martin & Loomis, 2013). Therefore, science teachers in the classroom need to make the environment where the students go through the process by using their *basic science process skills*. In 1960s, the Woods Hole Conference affected school science curriculum. At the conference, scientists and educators emphasized that schools need to have hands-on science to master the process of science.

Positions of professional societies

The National Science Teachers Association (NSTA, 2003) posed that "elementary science teachers should study content that is *balanced* among the main fields of science, and that they should study this material in the context of student-focused, laboratory-centered investigations."

The National Science Education Standards (NSES, 1996) states that:

Teachers of science must have a strong, broad base of scientific knowledge extensive enough for them to:

- Understand the nature of scientific inquiry, its central role in science, and how to use the skills and processes of scientific inquiry
- Understand the fundamental facts and concepts in major science disciplines
- Be able to make conceptual connections within and across science disciplines, as well

as mathematics, technology, and other school subjects

- Use scientific inquiry and ability when dealing with personal and societal Issues

Other professional societies and programs for science teachers include:

- National Science Teachers Association
- American Association for the Advancement of Science
- Goals 2000: Educate America Act
- National Association for the Education of Young Children

More professional societies and contemporary elementary science programs for science teachers can be found in Appendix A. All these organizations require science teachers to understand fundamental facts and concepts in major science disciplines but not all fields of science. The teachers do not have to know all of science. However, with basic process skills, the science teachers have their students explore the details of and make connections within and across science disciplines.

Summary

Through characteristics of science and history of science education, science is clearly defined as a *process* for producing knowledge, and science is always changing but continuously keeps the main clear and succinct ideas. Therefore, elementary science teachers need to understand fundamental facts and concepts of what they are going to teach, but they do not have to know everything in detail. The details are explored by the students with their basic skills: Observing, Measuring, Classifying, Predicting, Inferring, and Communicating.

The characteristics of science are: rejection of authority, consistency, and parsimony. Some of the first important events in the history of science were the launching of the Russian satellite (Sputnik) in 1957, the Hoods Hole Conference in 1959, and the emphasis of hands-on science in the classroom in 1960.

According to the NSTA (2003), "elementary science teachers should study content that is balanced among the main fields of science, and they should study this material in the context of student-focused, laboratory-centered investigations." In other words, teachers should have broad knowledge and basic skills in the main areas of science.

Assignments

Based on the information contained in this chapter, science teachers understand that their students need to improve basic process skills. This needs to be reflected in their lessons. Teachers must grasp the main ideas of what they are going to teach, and they need to make opportunities for students to improve their six basic process skills.

References

American Association for the Advancement of Science. (n.d.). *Science Process Skills*. Retrieved from: http://www.broward.k12.fl.us/corecurriculum/documents/pdf/science/Science_Process_Skills_Basic_Skills.pdf

Barba, R. (1998). *Science in the multicultural classroom, (2nd Ed)*. Needhan Heights, MA: Allyn & Bacon

Carin, A. A., Bass, J. E., & Contant, T. L.(2005). *Activities for teaching Science as Inquiry* (Sixth Edition). Upper Saddle River, NJ: Pearson Education Ltd.

Carin, Arthur A., Bass, Joel E., & Contant, Terry L.(2005). *Methods for teaching Science as Inquiry (Ninth Edition)*. Upper Saddle River, NJ: Pearson Education Ltd.

Currie, Nick. (2005). *Creativity and the Sputnik Shock*. Bloomberg L.P. Retrieved from: https://www.bloomberg.com/news/articles/2005-08-16/creativity-and-the-sputnik-shock

Different Atomic Models. (n.d.). Retrieved from: https://www.tutorvista.com/chemistry/different-atomic-models

Holt Science and Technology: Science Skills Worksheets. (2000). Holt, Rinehart, and Winston. Retrieved from: http://district.lindsay.k12.ca.us/view/8872.pdf

Kim, D., Suh, D., & Suh, J. (2009). Heungbu & Nolbu: Korean Two Brothers. Glpi Publisher.

Koch, Janice. (2005). Science Stories: Science Methods for Elementary and Middle School *Teachers (3rd Ed)*. Boston, MA: Houghton Mifflin

Martin, D. J. (2008). Elementary Science Methods: A Constructivist Approach, (Fifth Edition). Belmont, CA: Thomson/Wadsworth Inc.

Martin, D. & Loomis, K. (2013). Building Teachers: A Constructivist Approach to Introducing Education, (2nd Ed.). Belmont, CA: Cengage Learning.

National Research Council. (1996). *National Science Education Standards*, Washington, DC: National Academy Press.

National Science Teachers Association. (2003). *Standards for Science Teacher Preparation*, Arlington, VA: National Science Teachers Association.

Rutherford, F. J. & Ahlgren, A. (1991). *Science for All Americans*. Oxford University Press, USA; 2nd edition. http://www.project2061.org/publications/sfaa/online/ chap1.htm

Yoon, J. (2010). *Teaching Science & Environmental Education*, (Fall). Boston, MA: Pearson Custom Education

Tip of the Day 1: City of Angels

Hi future science teachers,

Have you seen an angel? Do you know what an angel looks like? I saw some angels. Last week when Mr. McCarthy tried to learn a new technology skill, I saw an angel in him. When Mrs. Stevenson laughed as loud as I could hear her in the hallway, I found an angel from her face.

But do you know the place where I can find many angels? That is your *classroom*! Whenever I visited classrooms, I met many angels. From your students who had curious eyes and were anxious about learning, I witnessed the angels. Then, how lucky you are! You are in heaven!

All the students around you are actually the angels! To meet the angels, make your students curious about and interested in your lessons!

Jiyoon

Inquiry

How Can We Be Inquiry-Based?

2

I had a beautiful white-dog named "Pluto." He liked to go to my backyard. Whenever I had him out, he started his own experiment. Usually, he checked his own smells everywhere. But whenever he smelled something different, he smelled again and looked at it. Sometimes, he stared fiercely at and even *licked* it to find out who it belonged to. I was always impressed by Pluto, who was such a wonderful scientist. I never taught him to do the experiments. However, he instinctively had the talent to do science. All I did was take him out and let him do his own experiment. Figure 2-1 shows Pluto doing his experiment in the backyard.

How will you make *inquiry* a priority in your science classroom? All of your students are naturally born to be scientists. All you need to do is develop an environment where your students can do their own exploration with freedom! This chapter defines inquiry and shows how to make science classrooms inquiry-based by comparing free-discovery and guided-inquiry instruction.

FIGURE 2-1 Pluto is smelling in the backyard

Definition of Inquiry

Children are born to be scientists. The astronomer Carl Sagan once said (National Academy Press, 1998, p. 1), "Everybody starts out as a scientist. Every child has the scientist's sense of wonder and awe." Children naturally wonder about many things around them and keep asking questions, thereby gaining new insights to help them understand and adapt to their surroundings (Yoon, J. & Onchwari, J. A., 2006). To continue their sense of wonder, teachers in schools need to prepare themselves to appropriately deal with children's curiosity.

However, with the traditional expository teaching method, it is not easy for teachers to provide young children opportunities to explore their world (Yoon, J. & Onchwari, J. A., 2006). The traditional expository teaching method gives teachers ideas about *what* to teach not *how* to teach by giving a big list of steps for a lesson activity. Instead, when they plan a lesson, teachers need to think about how they can make students inquire.

According to the dictionary (Mariam-Weber, 2010), inquiry is defined as a systematic investigation. A systematic investigation is an investigation that takes into account all aspects of problem, procedure, and results. Therefore, inquiry means systematically planned exploration. That includes systematic observation, measurement, and experiment, and the formulation, testing, and modification of hypotheses, which is the whole process, from question to outcome, of science. "Scientists use their background knowledge of principles, concepts, and theories, along with science process skills, to construct new explanations to allow them to understand the natural world. This is known as 'scientific inquiry'" (Lazoudis, A., Salmi, H., & Sofoklis, S, 2013, p. 32). The science in Chapter 1 was also defined as a process for producing knowledge. Science means a process and the whole process is inquiry:

Science is Process, Process is Inquiry, and Inquiry is Science. Based on this equation, we can say, "Science is Inquiry." However, if you want to delineate between science and inquiry, we can say science is a more abstract, mental process and inquiry is a more

actual, practical process for producing knowledge. Inquiry asks students to "do" science by following the whole process from questions to outcome.

When you teach science, you always need to make science classes inquiry based. That way, you can say you are teaching science. Science always goes with inquiry. Without inquiry in your class, you cannot say you are teaching science.

Guided Inquiry & Full Inquiry

When we consider an inquiry line (0% to 100%) of teaching methodology shown in Figure 2-2, there are three teaching methodologies available, based on the degree of inquiry (Martin, D. J., 2011, p. 194–195):

1. **No Inquiry** (**Expository Teaching**)—Totally teacher-centered, teacher as the controller of the class, teacher does the work and the students, who is or is not engaged in learning, is supposed to absorb the information.
2. **Guided Inquiry**—Teacher as the facilitator, students investigate topics established by the teacher in ways that are comfortable for the students and that also stimulate the students to ask and investigate additional questions suggested by the original explorations. It includes additional strategies, such as problem-based learning, differentiated instruction, and personalization.
3. **Full Inquiry** (**Free Discovery Teaching**)—Totally student-centered, students explore subjects of their own choosing in ways that are most comfortable to them. The teacher is the facilitator, and students are engaged in a variety of activities, such as investigating, experimenting, reading, writing, discussing, and exploring in other ways. This approach includes the project-based science strategy.

The free discovery methodology is located at the extreme right of the inquiry line. The following activity describes this methodology.

Science Activity 2-1 New Hairstyle

Activity Procedure:

1. Ask students to wear a new hairstyle when they come to the next class.
2. At the next class, talk with the students about the process of how they made their new hair styles, the reactions of friends and families, and their findings, like if they would like to have the new hairstyle in the future or not.

Throughout the discussion, the students would understand that they went through the whole process to find if the new hairstyle fit them or not. At the beginning, the students were curious about which hairstyle fits them (questions). To find the best hairstyle, they likely tried many hairstyles. Some of them might have looked at books, websites, or other resources. Once they decided a new hairstyle, they brought it to the class. Based on others' reactions to their new hairstyle, they may decide if they would keep the new hairstyles or not (findings: the product of science).

The free discovery methodology is the ideal teaching method that provides an environment where students work to construct meanings and discover new concepts by themselves. However, to develop the free discovery environment, teachers need to invest a lot of time and money. For example, the paper airplane competition takes about 40–50 minutes of class time. Besides, the teacher needs to provide various materials that the students may want to explore with to make their best airplanes.

Compared with the free discovery methodology, the guided inquiry teaching methodology provides enough structure to eliminate the sense of wandering that students may get in free discovery approaches. Students are given parameters by which to start their inquiries and such necessary constraints as time, group size, materials, and so on. The class is manageable, and the curriculum is covered. Guided inquiry also uses an inquiry constructivist approach in which science content is used as a vehicle for mastery of the process.

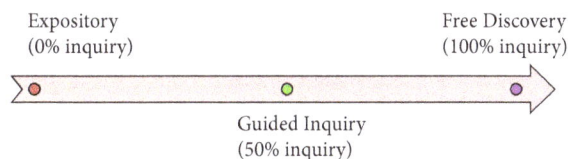

FIGURE 2-2 Inquiry Line for Teaching Science

The following activity provides the opportunity to move safely toward guided inquiry.

Problem-based learning (PBL) is a form of guided inquiry that centers on finding the solutions to problems. In PBL, the teacher sets up problems that are based on the curriculum. The teacher then involves the students in developing one or more viable solutions to the problems by posing essential questions, providing for constructive investigations where students explore and construct their own understanding and solutions and encouraging an appropriate degree of student autonomy (Fallik, Eylon, & Rosenfeld, 2008). The learning experiences provide students with real-life problems that form the basis and directions of their explorations. Examples of typical problems include global warming, ownership of water flowing in rivers, alternative energy sources, the safe production of food, and the sustainability of America. The goal is for students to become better problem solvers, not necessarily to find definitive solutions to the problems.

Project-based science (PBS) is similar to problem-based learning except that the focus is on projects that can actually be implemented locally rather than problems that are larger in scope. It is a student-centered approach (free discovery) in which students pose and answer research questions that are relevant to their own lives and communities (Colley, 2008).

Science Activity 2-2 Paper Airplane Competition

Activity Procedure:

1. The teacher asks the students to develop their own paper airplanes that can fly further than others'. Further refers to distance, not height, since height is difficult to measure correctly.

2. The students can research using websites or books to learn how to fold paper airplanes that can fly furthest (research time should last about 30 minutes).

3. Once they select a model for their best airplane, they can try flying their selected airplanes in groups and pick the best one per group (for about 10 minutes).

4. The groups go outside of the classroom (or hallway) and fly the best airplane among group members'. The airplane that flies furthest is the winner.

5. Once they come back to the classroom, they listen to the winner talk about how he or she made the airplane fly further. The students may understand concepts of aerodynamics, air resistance, and air pressure.

In this activity, the teacher just throws out a prompt at the beginning and students explore how to make the best airplane that flies further than others' and find new concepts for themselves.

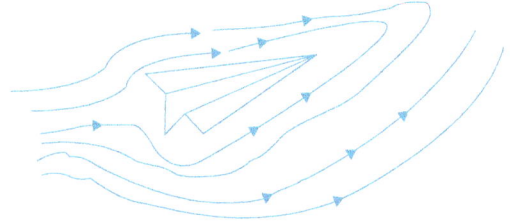

FIGURE 2-3 Aerodynamics of flying a paper airplane

Science Activity 2-3 Guided inquiry to learn about a Llama

Activity Procedure:

1. This activity starts with my story: "My husband took a picture of me at a local zoo, and I found one animal covered more than half of my picture (see Figure 2-4). I was wondering what this animal was." The students should guess what this animal is.

FIGURE 2-4 A picture of me at a zoo

2. After the students guess if she is a llama, donkey, or camel, show the students the next picture (Figure 2-5) to present the answer: Llama.

FIGURE 2-5 This picture shows the answer

3. Ask the students the following questions, and then, have them search websites to find the answers:

a. Describe this animal.
b. Where else can you find this animal?
c. How is this animal different from a camel?
d. How can you design a house for this animal?
e. Is it good or bad to give this animal a sausage?

Through these questions, the students explore ideas about llamas. They understand where the llama lives, how they are different from camels, and what kinds of houses they live in. At the end of the questions, the students can evaluate everything about llamas. The questions prepared by the teacher can guide the students to find everything out about llamas. This kind of inquiry saves times and money both for the teacher and students learning the new concepts.

Questioning

Inquiry-oriented teachers seldom tell, but often question. By asking questions, students move from observation to the development of predictions. Through questions given by teachers, children make their discoveries and use their minds. Because of its potential for stimulating thinking and learning, questioning remains one of the most influential teaching strategies in science classrooms. Therefore, teachers must formulate questions in lessons that will challenge and, at the same time, guide children (Yoon, J. & Onchwari, J. A., 2006).

When science teachers develop questions, Bloom's taxonomy can help compose questions on different levels of thinking. This taxonomy ranges from lower to higher levels of cognitive thinking. There are six levels (Forehand, 2005):

1. Knowledge (recall)
2. Comprehension (understanding)
3. Application (using knowledge in a new situation)
4. Analysis (breaking things down, critical thinking)
5. Synthesis (putting things together, creative thinking)
6. Evaluation (judgment)

The Evaluation and the Synthesis can be switched with each other, depending on the views of researchers and educators. So the Synthesis is possibly the last level of cognitive thinking. There are many researchers who provide techniques and examples to develop questions based on Bloom's taxonomy. Appendix B shows helpful sites for developing questioning. The following are the examples of question stems in each thinking level of taxonomy (Richli, 2006):

KNOWLEDGE

Useful Verbs	Sample questions
Define Describe Discuss, Explain Express, Identify List, Name Recall	Can you name the . . . ? Define what happened at . . . ? Who spoke to . . . ? Can you recall . . . ? What is . . . ?

COMPREHENSION

Useful Verbs	Sample questions
Abstract, Categorize, Clarify, Classify, Compare, Contrast, Exemplify, Explain, Generalize, Paraphrase, Represent, Summarize, Translate, Illustrate, Interpret, Match, Model	Can you explain in your own words . . . ? Can you classify . . . ? What do you think . . . ? What was the main idea . . . ? Can you compare each other? Can you provide a model of what you mean . . . ?

APPLICATION

Useful Verbs	Sample questions
Carry out Execute Implement Use	Do you know another instance where . . . ? Can you apply the method used to some experience of your own . . . ? What questions would you ask of . . . ? From the information given, can you develop a set of instructions about . . . ? Would this information be useful if you had a . . . ?

ANALYSIS

Useful Verbs	Sample questions
Analyse, Deconstruct Differentiate Distinguish Find coherence Focus, Integrate Organize, Outline Parse, Select Structure	Can you distinguish between . . . ? What were some of the motives behind . . . ? What was the turning point in the game? Can you outline the event of . . . ?

SYNTHESIS

Useful Verbs	Sample questions
Analyse, Deconstruct Differentiate Distinguish Find coherence Focus, Integrate Organize, Outline Parse, Select Structure	Can you design a . . . to . . . ? Can you generate a possible solution to . . . ? Construct your own way to deal with . . . ? What would happen if . . . ? Can you plan a new project for . . . ? Can you write a new recipe for . . . ? Can you develop a proposal which would . . . ?

EVALUATION

Useful Verbs	Sample questions
Assess, Argue Choose Debate, Decide Determine Evaluate Judge, Justify Verify, Rate Recommend Prioritise	Is there a better solution to . . . ? Justify the value of . . . Can you determine your position about . . . ? Do you think . . . is a good or a bad thing? How would you rate . . . ? What changes to . . . would you recommend? Do you believe? Are you a . . . person? How would you feel if . . . ? What do you think about . . . ?

John Dewey (1910, 30) said, "Wonder is the mother of all science." The curious mind is constantly alert and exploring, seeking materials for thought. "Eagerness for experience, for new and varied contacts, is found where wonder is found" (Dewey, 1910). To inspire curiosity in students, teachers need to prepare questions appropriate to the students' thinking levels before teaching science.

It is important to consider stressing high-level questions and devising a variety of instructional objectives that balance low-level memory questions with preplanned high-level divergent questions. Unfortunately, research studies on the use of questions as a learning strategy have found that K–5 Teachers persistently asked questions that primarily require children to recall knowledge and information (Trowbridge et al., 2004). It is required that teachers use both low and high cognitive level questions to propel science learning.

For a child who is able to think at a level, he or she must possess the necessary information and skills. "He or she must know the facts, understand them, and be able to apply them to different and unique situations" (Yoon, J. & Onchwari, J. A., 2006). Having that ability with a certain collection of facts, the child can then proceed to analyze, synthesize, and evaluate with them. This thinking must be fostered through the next higher-level questions developed by teachers. This is called by Vygotsky's Zone of Proximal Development (ZPD).

Teachers check with their students to see what thinking levels they have and prepare questions both in the current thinking level and a little bit higher (or next higher) level. For example, if the students are in the Knowledge level, then the teacher needs to prepare questions both in Knowledge and Comprehension level. Teachers use those questions where less competent students develop with help from the teacher—within the zone of proximal development. "Vygotsky believed when a student is in ZPD for a particular task, providing appropriate assistance will give the student enough of a boost to achieve the task" (Yoon, J. & Onchwari, J. A., 2006; Silver, 2011). With the questions in a little bit higher level, teachers can improve their students' thinking.

The famous science educator, Gasset (1963) tells clearly what teachers need to do for inquiry-based teaching:

> He who wishes to teach us a truth should not tell it to us, but simply suggest it with a brief gesture, a gesture which starts an ideal trajectory in the air along which we glide until we find ourselves at the feet of the new truth.

In the guided inquiry–based classroom, teachers are invisible, work as guiders, and do not tell; they simply suggest with a simple question, until the students find the truth. Also, teachers need to remember that all the guided inquiry must be done slowly to give the students time to explore and reconstruct.

The following activity introduces another guided-inquiry teaching methodology that teachers can do with diverse students in science classrooms.

Science Activity 2-4 Being a scientist

Activity Procedure:

1. Divide the class into two (or more, depending on how many scientists the class will research) groups. In this class, there are two scientists, Galileo and King Sejong. King Sejong is from Korea. If your students are from other cultures, check with them to find famous scientists from their own cultures.

2. Provide websites and books for each of the groups to find information about their scientists.

3. Ask students to post their findings on an electronic discussion board or board in the classroom and set aside a time to share their findings.

4. After sharing the information about the scientists, the students actually develop the most remarkable inventions that their scientists made. The Galileo group will do the Egg Drop Project (gravity), and the King Sejong group will do the "Invention contest for measurement tools."

5. Hold a musical with the Galileo and King Sejong groups. Before the musical performance, groups get together and develop scripts that include information about each scientist and imagine if the two scientists lived at the same time and met in real life.

This activity may take more than four class hours. Teachers choose a month for the scientists and have students research their scientists' lives and inventions for the whole month. Through this activity, the students actually become the historically famous scientists and construct their scientific literature from their lives.

Summary

Inquiry is the systematic investigation that includes the whole *process* of doing science. Therefore, we can say inquiry is science. All science classes are required to be fully inquiry based. In other words, without inquiry, you cannot say you teach science. Free discovery is an ideal inquiry-based teaching methodology. However, guided-inquiry teaching methodology is more recommended in science classrooms, because of its effectiveness in real school situations. For the guided inquiry, teachers need to develop questions in advance of teaching. Bloom's taxonomy helps to compose the questions in different thinking levels. The questions need to be in the current and a little bit higher thinking level for improving students thinking within a zone of proximal development.

Assignment

To make inquiry-based science classrooms, teachers are required to have skills to develop questions according to students' different thinking levels. Choose a science concept and a grade level to teach and develop at least six questions (at least one question per Bloom's thinking level). For example, when you want to teach the water cycle, you can develop a question for the Knowledge level, like "What is water cycle?" For the Comprehension level, you can have the question, "Describe precipitation in your own words." Even though your students are in lower grade level, you may have some students with a higher level of thinking. So, to prepare for all kinds of situations in classrooms, please develop questions for all the thinking levels.

References

Colley, K. (2008). "Project-based Science Instruction: A Primer, An Introduction and Learning Cycle for Implementing Project-Based Science." *Science Teacher, 75*(8), 23–28.

Richli, L. (2006). Blueprint for Learning: constructing College Courses to Facilitate, Assess, and Document Learning, 45–51.

Dewey, J. (1910). *The Problem of Training Thought.* Boston: D.C. Heath & Co.

Fallik, O., Eylon, B., & Rosenfeld, S. (2008). "Motivating teachers to enact free-choice PBL in science and technology (PBLSAT): Effects of a professional development model." *Journal of Science Teacher Education, 19,* 565–591.

Forehand, M. (2005). Bloom's taxonomy: Original and revised. In M. Orey (Ed.), Emerging perspectives on learning, teaching,

and technology. Retrieved March 1st 2015, from http://epltt.coe.
uga.edu/index.php?title=Bloom%27s_Taxonomy.

Gasset, J. O. (1963). *Meditations on Quixote*. W. W. Norton &
Company.

Lazoudis, A., Hannu, S., & Sofoklis, S. (2013). *Augmented Reality in
Education: Proceedings of the "Science Center To Go" Workshops*.
Ellinogermaniki Agogi. 32.

Martin, D. J. (2011) *Elementary Science Methods: A Constructivist
Approach*. Wadsworth Publishing. 194-195.

National Academy Press. (1998). *Every Child a Scientist: Achieving
Scientific Literacy for All*. Washington D.C. VA: National
Academy Press.

Silver, D. (2011). Using the Zone Help Reach Every Learner. *Kappa
Delta Pi Record, 47*, 28–31.

Yoon, J & Onchwari, J. A. (2006) "Teaching Young Children
Science: Three Key Points." *Early Childhood Education Journal,
33*(6). 419-423.

Tip of the Day 2

Hi future science teachers,

Today, I want to share a picture of two jars that I found from a website. On each jar in the picture, there is a sign that shows the date when the French fries were stored in the jar. A 5th -grade Polish girl was so curious about how the French fries would be five years later. So, she decided to keep the French fries from two different fast food restaurants in two different jars and then took pictures of them five years later. Guess what happened? One jar had rotten French fries, but the other jar had well-formed French fries, even after five years. After this experiment, she understood why people kept saying that French fries at some fast food restaurants were not good for your health. They are using a lot of food preservation that is not good nor healthy.

I want you make your science classroom fully inquiry based so that your students always inquire any time and any place, just like this girl.

Jiyoon

Lesson Plan

How Do We Plan a Science Lesson?

3

My husband and I like to go out and play golf. I enjoy not only playing golf itself, but also science on the golf course. I often saw many tortoises wandering around between ponds, heard woodpeckers pecking wood, and smelled all different kinds of wild flowers (Life Science). Also, I learned to read lies on the green and understand the relationship between the green and the weather (Earth Science). Further, I knew that the distance that my golf ball went was associated with the orbit of my swing and the angles of my grip (Physical Science). When I categorize the science concepts that I learned on the golf course, I realize they were components of content standards to teach science: life science, physical science, and earth science.

When science teachers start teaching science, they need to cover all the sciences that students will face in their daily lives. Science standards help teachers plan science lessons, making sure they are covering all the required science contents. This chapter provides teachers with ideas of how to plan science lessons based on national standards and the 5E learning cycle, including questions. To challenge students' misconceptions, discrepant events also need to be introduced at the beginning of lessons.

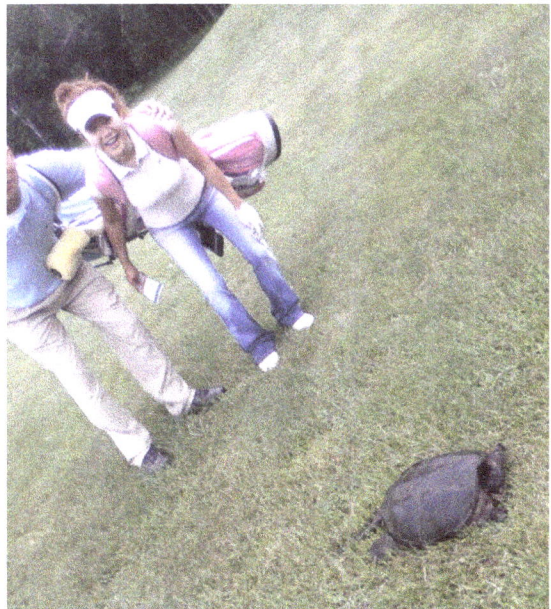

FIGURE 3-1 Tortoise on the golf course

Science Activity 3-1 Pepper and Water Science Magic Trick
(Rajil, T. L., 2014)

Materials: black pepper, water, dishwashing liquid, plate or bowl

Activity Procedure:

1. Pour water into a plate or bowl.
2. Shake some pepper onto the water.
3. If you dip your finger into the pepper and water, nothing much happens.
4. If you put a drop of dishwashing liquid on your finger and then dip it into the pepper and water the pepper will rush to the outer edges of the dish. If you are doing this as a trick, then you might have one finger that is clean and another finger that you dipped in detergent before performing the trick.

How it works:

When the detergent is added to the water, the surface tension of the water is lowered (in other words, it made the bonding of the water molecules weak). When the surface tension is lowered, the water wants to spread out. As the water flattens on the dish, the pepper that is floating on top of the water is carried to the outer edge of the plate as if by magic.

Through this science magic show, the students can understand the concepts of water surface tension and molecular bonding. Therefore, the science magic show can take students to meaningful inquiry and the scientific principles through an engagement.

In addition to learning science concepts, science magic shows improve students' observation, inference, and prediction skills. When students observe the magic shows, they are motivated and have a chance to think critically about how the events happened. In the following science magic show, students can make observations and inferences about what happens.

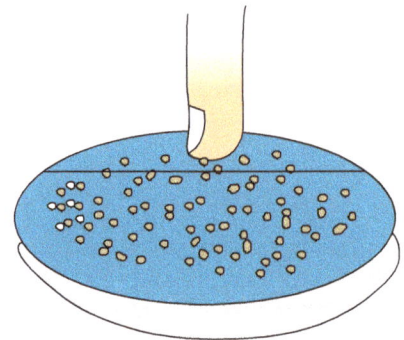

FIGURE 3-2 Pepper and water science magic

Discrepant Events

As a science teacher, it is important to gain students' interest to promote their learning. But it is a challenging for the teacher to spark the curiosity to truly engage students in the processes of "doing science."

One often-used method of motivation is the demonstration of science in action, such as "discrepant events." "They are 'discrepant,' because something occurs that is unexpected. A discrepancy causes students to wonder 'How did that happen?' and to want those questions answered" (Crawford, T., 2003).

Discrepant events influence equilibration and the self-regulatory process, according to the Piagetian theory of intellectual development. Situations that are contrary to what a student expects cause him or her to wonder what is taking place, resulting in cognitive disequilibrium. With proper guidance, the student will attempt to figure out the discrepancy and search for a suitable explanation for the situation. When a student arrives at a plausible explanation for a discrepant event, he or she will establish cognitive equilibrium at a new level. The student is now better mentally equipped to approach new situations that cause curiosity and puzzlement (Koballa, T., n.d., citing Piaget, 1971).

One of the discrepant events that teachers can demonstrate in the science classroom is a science magic show. The science magic show makes students aware of the inadequacies of their own explanations by exposing them to a demonstration that directly contradicts their ideas, therefore creating cognitive conflict.

Science Activity 3.2 Upside Down Water Trick ("Physics Over the Sink," n.d.)

Materials: transparent cup, index card, and water

Activity Procedure:
1. Fill the transparent cup with water.
2. Place the index card over the mouth of the transparent cup.
3. Hold the index card in place, slowly turn the transparent cup upside down, and then remove your hand from the index card. What happens?

How it works:
The index card stays in place, because air pressure is pushing up against the index card. The pressure pushing up is more than the weight of the water pushing down from the inside of the transparent cup, therefore the index card stays in place. This is called air-pressure.

The following science magic show will stimulate students' interest in science and spur their creativity.

To find out how water can be upside down, students pay attention to the magic demonstration making observations and inferences about what happened. The students are driven to inquiring about science like a magic. Science magic shows encourage students to understand and apply the science process skills of observation and inference as engagement.

Misconception

Science magic shows are developed based on children's misconceptions. The science magic shows are powerful ways to stimulate interest, motivate students to challenge their science misconceptions, and promote higher-order thinking skills.

Misconceptions are referred to as preconceived notions (Smith, I.E., al. 1993), intuitive beliefs (McCloskey, 1983), naive theories (Caramazzaet.al, 1981), mixed conceptions, or conceptual misunderstandings.

In science, there are cases in which something a [child] knows and believes does not match what is known to be scientifically correct. When they are simply told they are wrong, they often have a hard time giving up their misconceptions, especially if they have had the misconception for a long time.

What is especially concerning about misconceptions is that [students] continue to build knowledge on their current understandings. Possessing misconceptions can have serious impacts on [student] learning ("Curriculum Package," 2010, p. 105).

Science teachers have to challenge children's thinking and give them new perspectives. Appendix C shows children's misconceptions about science.

Science teachers need to spot the conflicts that arise between children's preconceptions and the more formal scientific concept, using discrepant events like magic shows. The water pump magic is to clarify the difference between misconception and scientific conceptions about states of matter (solid, liquid, gas), thermal expansion, air pressure, and vacuums.

Learning Cycle

The big difference between a science magic teacher and a professional magician can be found after their tricks are performed. The magician expects a nice round of applause from the audience and then introduces his or her next magic. However, the science teacher poses the question to the audience, "Do you know why it happened?" The applause is replaced with screams of "YES!" as the science magic teacher proceeds to share the science behind the trick (Smith, 2006).

Science Activity 3-3 Water Pump Magic trick ("Great Candle Mystery," n.d.)

Materials: two candles, matches, two glass jars large enough to hold the candles, two Petri Dishes, water, a pen, an adult, and a stop watch.

Caution! *This activity involves using matches and fire. For any activity involving fire, an adult's presence is required, as well as water, a fire extinguisher, and any other safety equipment. Take care that students only use fire in a safe place.*

Activity Procedure:

1. Put a candle in the middle of each of the two Petri Dishes.
2. Pour a couple of centimeters of water into each of the Petri Dishes.
3. Light the candles.
4. Carefully place a jar over one of the candles.
5. Time how long it takes between putting the jar over the candle and when the candle goes out. After the candle has gone out, some water will be sucked into the jar.
6. Mark the water level on the jar.
7. Now, place a jar over the other one of the candles.
8. Compare the time it takes for the candles to go out and the water level this time with the previous time.

If the jar fits perfectly against the Petri Dish, you may find the water can't get in. In this case, the jar will not fill with water, but will be held against the bowl like a suction cup. To keep this from happening, place some small coins around the candles so the jar rests on them with a small gap to allow the water through.

How it works:

Oxygen makes up about one-fifth of the air. When the candles burn, they use oxygen from the air. When there is no more oxygen in the air, they go out. This theory often explains how it works. However, this theory cannot provide the reason why the second jar has more water sucked up than the first jar. The theory is a misconception that students might have.

The real reason why the water sucks up is related to air expansion. When the jar covered the second candle, the fire heated enough of the air inside the jar. As it was heated, the air expanded. Once the candle went out, the air in the jar cooled down and contracted. The water was then pushed in by the surrounding air as the gas pressure inside the jar was lower due to there being fewer gas molecules inside the jar than when the whole procedure started.

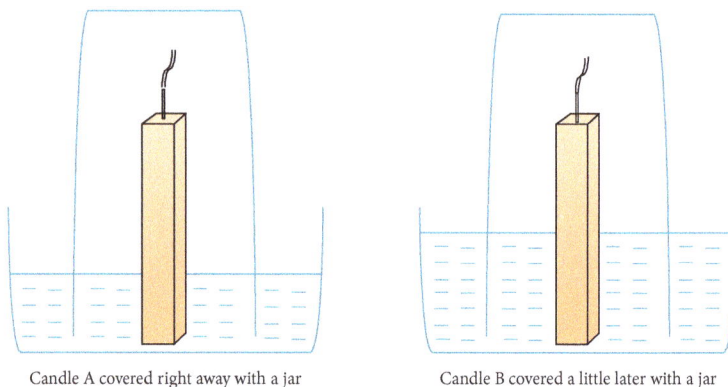

Candle A covered right away with a jar   Candle B covered a little later with a jar

FIGURE 3-3 Water pump magic

The learning cycle is an effective instructional method to learn about science principles and concepts behind the magic secret. Progressing from the questions generated by discrepant event demonstrations, the learning cycle forms and applies concepts to new situations.

The 5E learning cycle was developed by the Biological Sciences Curriculum Study group (Biological Sciences Curriculum Study, 1989). There are five phases in the 5E learning cycle (Carin et al., 2003):

Engagement: Teachers engage students in questions about objects, organisms, and events in the environment and probe background knowledge and conceptions

Exploration: Teachers have students plan and conduct investigations to gather evidence to answer the questions

Explanation: Building on students' explorations and explanations, teachers formally present labels, concepts, and principles; students, guided by the teachers, use new knowledge to construct scientific explanations and answer initiating questions

Elaboration: Teachers provide opportunities for students to apply new understandings to problems

Evaluation: Teachers use formative and authentic assessment means to assess students' new knowledge, understanding, appreciation, and abilities in each of 5E phases.

The 5E learning cycle consistently helps students explore science. The elaboration is not the end of the cycle but the trigger of the next learning cycle. Evaluation occurs in all four Es (Engagement, Exploration, Explanation, and Elaboration) of the learning cycle, ensuring every student's understanding in each E.

When teachers plan science lessons and units around the five phases of the learning cycle, students can move from concrete experiences to the development of understanding to the application of principles.

Studying surface tension is a way to introduce students to a physical science lesson on the properties of objects and materials. This lesson follows the 5E learning cycle format and is appropriate for grades three through six.

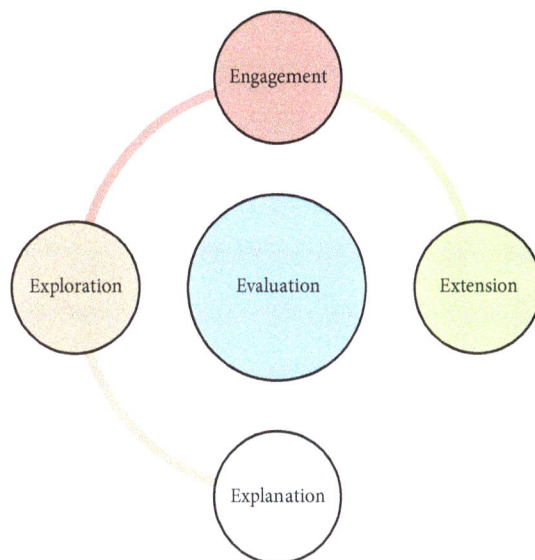

FIGURE 3-4 5E Learning Cycle Model

Science Activity 3-4 Sticky Water ("Water Chemistry Activities," n.d.)

Materials: plastic cups or beakers to hold water, coins (pennies, nickels, dimes), paper towels, and worksheet

Activity Procedure:

1. Engagement: Show a photo of a water strider, and ask students "How can this water strider walk on water?"

   a. *The answer is surface tension. Even though the students cannot draw the concept of surface tension, they understand the water skin is strong enough to withstand the weight of the water strider.*

2. Exploration: Divide students into two teams. Give each pair a coin, an eyedropper, a cup or beaker of water, and paper towels.

   a. Have the teams predict how many drops of water they will be able to put on the coin.

   b. Allow the students to slowly begin to place drops of water on the coin. Tell the students to add the drops one drop at a time for better results. Students will count the drops and continue to add drops until the surface tension breaks, the water drop collapses, and spills over the side of the coin.

   c. Have the students record the number of drops they were able to successfully place on the coin before the water drop collapsed.

   d. Allow the other student on the team to repeat the activity.

3. Explanation: Students report their team's data to the class. Share the reasons why so many water drops can be piled up on the coin. After the students' discussion, the teacher introduces the concept of water surface tension.

How it works:

The dome is a result of a characteristic of the water molecule called surface tension. A water molecule is made up of three atoms: two hydrogen atoms and one oxygen atom. What is interesting about these atoms is that the hydrogen atoms have a positive charge and the oxygen atom has a negative charge. In a drop of water, there are millions of water molecules. When so many water molecules are put together, the negative oxygen atoms are attracted to the positive hydrogen atoms and a very weak bond is formed that is called a hydrogen bond. It is the hydrogen bond that causes the molecules to stick together, resulting in surface tension. And surface tension forms the dome shape.

4. Elaboration: Try testing the surface tension of liquids other than water at home. Experiment with vinegar, milk, oil, soda, shampoo, and dish liquid. Remember to use the same size coin for each experiment. The liquid that piles up the most on the coin without spilling over the edge has the strongest surface tension. More elaboration activities of water surface tension can be found through the Internet.

5. Evaluation: Students' learning is evaluated through their answers to the questions and experiment worksheet (see the following worksheet).

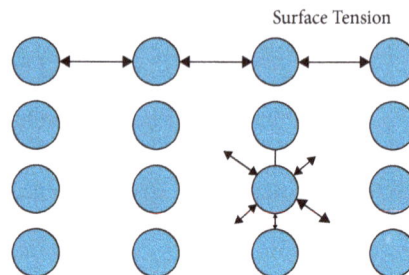

FIGURE 3-5 Water surface tension

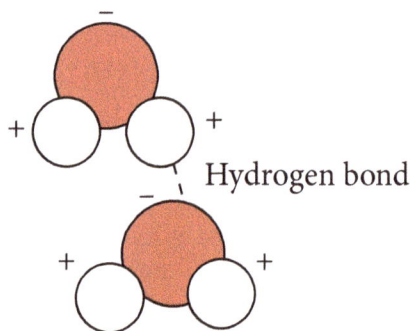

FIGURE 3-6 Hydrogen bond in water

Sticky Water Worksheet

Name: _____

Prediction: I will be able to put _____ drops of water on the coin.

Directions: Slowly begin to place drops of water on the coin, add the drops one drop at a time for better results. Count the drops until the water drop collapses and spills over the side on the coin. Record the number of drops below. Allow the other team to repeat the activity. Report your team's data to the class.

Number of Drops: Team Member One: _____ Team Member Two: _____

Teams	Penny		Nickel		Dime	
	Heads	**Tails**	**Heads**	**Tails**	**Heads**	**Tails**
Team 1	drops	drops	drops	drops	drops	drops
Team 2						
Team 3						
Team 4						
Team 5						
Team 6						
Team 7						
Team 8						
Team 9						
Team 10						
Team 11						
Team 12						
Team 13						
Team 14						
Team 15						

Source: ("Sticky Water Worksheet," 2003)

Questions

Inquiry-based science classrooms are always full of questions. In Chapter 2, questioning is one of the ways to promote students to explore science. The following are examples of questions to ask students in order to teach plant growth. These questions encompass the full spectrum of Bloom's taxonomy (Martin, 2003).

1. Name the parts of a plant (Knowledge).
2. What part of the plant is this? (Teacher points to the stem.) How do you know? (Comprehension)
3. What are the leaves on this plant? (Teacher shows a plant children have not studied, such as an evergreen.) (Application)
4. If a plant were to lose all its roots, would it be able to live? (Analysis)

TABLE 3-1 Lesson plan integrated with 5Es and Questions.

Grade K–5
Topic Question How can you make ice cream?
Objectives 1. Students will make ice cream using basic household ingredients and compare its taste with store-bought ice cream. 2. Students will observe the chemical reactions that take place when making the ice cream.
Materials milk, cream, or half & half vanilla extract or chocolate syrup sugar, ice, salt baggies (Ziploc) (large and small sizes) some newspaper

Phases of the 5E Model	Questions
Engage. Ask questions about objects, organisms, and events in the environment.	1. Have you ever eaten ice cream? (Knowledge) 2. Describe the taste of the ice cream? (Comprehension) 3. Name the materials you think the ice cream was made from? (Knowledge) 4. How do you think is ice cream made? (Analysis)
Explore. Ask questions to gather evidence to answer the question posed.	1. How can you make ice cream with the materials? (Synthesis) 2. Find several ways to make ice cream (Synthesis). *Bring the materials for making ice cream. Each student puts the ingredients into a sandwich-size Ziploc bag. Put three or four of the student's bags into a large Ziploc bag that is half-filled with ice and salt and shake the bag for about 5 mins. Continue shaking for several more minutes, if necessary.*
Explain. Ask questions to use new knowledge and observable evidence to construct scientific explanations and answer initiating questions.	1. Why are the liquid ingredients turning into ice cream? (Analysis) 2. How long does it take to turn the ingredients into ice cream? (Knowledge) 3. What does the salt do in the ice cream–making process? (Analysis)
Elaborate. Ask questions to apply new understandings to new problems.	1. Is it possible to make different flavors of ice cream? (Applications) 2. Can you make an ice cream by excluding one of the ingredients? (Synthesis) *Students will have a fieldtrip to an ice cream factory and will find out how ice cream is made in the factory.*
Evaluate. Ask questions to assess developing understanding and inquiry skills.	*Use continuous (formative) assessment; assess performance on the activity sheet (for older children), and/or oral explanations and predictions.*

5. Suppose you live on a far-away planet. Draw a plant that would grow on that planet. (Synthesis)
6. What do plants need in order to live? How would you prove that? (Synthesis)
7. Do the rest of you agree that this plant would grow on the planet he or she has described? (Evaluation)
8. From our experiment, did we prove that plants need food, light, and water in order to live? (Evaluation)

When teachers plan their lessons, they need to not only think about incorporating inquiry into the plan, but also the questions they would wish to ask children. The way teachers ask questions, and the amount of time they provide their students for inquiry influences children learning. Table 3-1 is an example of an inquiry-based science lesson. The lesson is based on the 5E model of instruction integrated with questions.

With the 5E learning cycle model, teachers can prepare themselves in advance to provide students with chances to explore specific concepts and explanations and the students can learn science by an inquiry-based step-by-step fashion than by the traditional expository approach. Appendix D gives more sites for developing science-lessons based on the 5E learning cycle model.

Summary

In developing science lessons, teachers need to consider national standards to see if they are following national expectations and teaching required content. Especially, their topics for the lessons need to be aligned with the national and state content standards. The 5E learning cycle model supports an inquiry-based learning environment, following five phases and promoting student learning by using inquiry and questions. The five phases are composed of Engagement, Exploration, Explanation, Extension/Elaboration, and Evaluation. After the Extension, the engagement of the next learning cycle starts. Each of the five phases needs to be evaluated throughout the lesson. For the engagement phase, teachers in the science classroom use discrepant events that make students dis-equilibrate cognitively. One of the excellent discrepant events is a science magic show. The science magic show develops the cognitive disequilibrated learning environment and also improves observation and inference skills among students. In each of the five phases, teachers need to embed questions to lead students from observation to the development of inference to inquiry. The questions are developed based on Bloom's taxonomy. In planning science lessons, teachers utilize a higher order level of questions to propel students' critical thinking.

Assignment

Science teachers practice developing their science lessons based on the 5E learning cycle model, including questions. The lesson topic must match with the national/state content standards, along with the grade level that the teachers will teach. The questions in each of the five phases apply the Bloom's taxonomy.

References

Biological Sciences Curriculum Study. (1989). New designs for elementary school science and health: A cooperative project of Biological Sciences Curriculum Study (BSCS) and International Business Machines (IBM). Dubuque, IA: Kendall/Hunt.

Boudourides, M. A (2003). Constructivism, education, science and technology. Canadian Journal of Learning and Technology, 29(3), 14–23.

Carin, A. A., Bass, J. E., & Contant, T. L. (2003). Methods for Teaching Science As Inquiry, 9, 111–123.

Caramazza, A., McCloskey, M., & Green, B. (1981). Naive beliefs in sophisticated subjects: misconceptions about trajectories of objects. Cognition, 9:117–123.

Center for Science, Mathematics, and Engineering Education (CSMEE). (1998). Every Child a Scientist: Achieving Scientific Literacy for All. Washington, DC: The National Academies Press.

Chaille, C., & Britain, L. (2003). They young child as scientist (3rd ed.). New York: Allyn & Bacon.

Crawford, T. (2003). "From Magic Show to Meaningful Science." Science Scope, 27(1). 36-39.

Curriculum Package 2010-2011: Chemistry. (2010). Twin Rivers Unified School District. 105. Retrieved from: http://marric.us/files/TRUSD_Chemistry_Curriculum_Package_2010-2011.pdf

Foster, Geral W. (2003). Elementary Mathematics and Science Methods: Inquiry Teaching and Learning. Belmont, CA: Wadsworth Press. 65–79.

Great Candle Mystery. (n.d.). Retrieved from: http://www.giant-classroom.com.au/program.aspx?ProgramId=8&ProgramLink-Id=54

Harlen, W. (2001). Research in primary science education. Journal of Biological Education, 35(2), 61–65.

Harlan, J. D., & Rivkin, M. S. (2004). Science experiences for the early childhood years: An integrated affective approach (8th ed.) Upper Saddle River, NJ: Pearson.

Hamilton, R., & & Ghatala, E. (1994). Learning and instruction. New York: McGraw.

Keil, F. C., & Wilson, R. A. (2000). Explaining Explanation. In F. C. Keil & R. A. Wilson (Eds.), Explanation and cognition. Cambridge, MA: The MIT Press, 1–18.

Koballa, T. (n.d.). *The Motivational Power of Science Discrepant Events.* Retrieved from: https://www.scribd.com/document/137151003/The-Motivational-Power-of-Science-Discrepant-Events-docx

Llewellyn, D. (2010). Inquire within: Implementing Smith, I.E., Blakeslee, T.D. & Anderson, C.W. (1993). Teaching strategies associated with conceptual change learning in science. Journal of Research in science teaching, 30(2), 111–26.

Lorsbach, A., & Tobin, K. (2004). "Constructivism as a referent for science teaching." Retrieved on November 12, 2004 from http://www.exploratorium.edu/IFI/ resources/research/constructivism.html

Lowery, L. F. (Ed.). (1997). Pathways to the science standards: Elementary Edition. Arlington, VA: National Science Teachers Association.

Markezich, A. (1996). Learning windows and the child's brain. Super Kids Educational Software Review. Knowledge Share LLC.

Martin, D. J. (2003). Elementary Science Methods: A Constructivist Approach. Belmont, CA: Wadsworth, Thomson Learning, Inc., 234–235.

McCloskey, M. (1983). "Intuitive physics." Scientific American, 248(4), 114–122.

National Research Council. (1996). National Science Education Standards. Washington DC: National Academy Press.

National Research Council. (2013). "Next General Science Standards for States, By States." Reviewed April 20, 2015 from http://www.nextgenscience.org/next-generation- science-standards

National Science Teachers Association. (1998). "NSTA Position Statement: The National Science Education Standards." Retrieved on July 1, 2011 from http://www.nsta.org/about/positions/standards.aspx

Physics Over the Sink: Water Glass Magic. (n.d.). Retrieved from: http://www.physicscentral.com/experiment/physicsathome/magicwaterglass.cfm

Piaget, J. (1971). Biology and knowledge. Chicago: University of Chicago Press.

Rajil, T. L. (2014). Water-Pepper, Soap Oil Experiment. Retrieved from: http://www.trendshere.com/2014/06/water-pepper-soap-oil-experiment.html

Smith, A.V., King, J.A., Orr-Weaver, T.L. (1993). "Identification of genomic regions required for DNA replication during Drosophila embryogenesis." Genetics, 135(3), 817–829.

Smith, L. (September, 2006). M-U-M. Society of American Magicians. Retrieved from: https://www.stevespanglerscience.com/2006/08/31/mad-about-science-magic/

Sticky Water Worksheet. (2003). Stevens Institute of Technology, CIESE. Retrieved from: http://hrsbstaff.ednet.ns.ca/jcarroll/millwood/oceans%2011/structure%20and%20motion/stickywaterworksheet.doc

Trowbridge, Leslie W., Bybee, Rodger W., & Powell, Janet C. (2004). Teaching Secondary School Science: Strategies for Developing Scientific Literacy, Arlington, VA: NSTA Press, 112–123.

Water Chemistry Activities. (n.d.). The Global Water Sampling Project. Retrieved from: http://www.k12science.org/curriculum/waterproj/waterchemistryactivities/

Tip of the Day 3: Be a Savvy Customer

Hi future science teachers,

What is your plan for this weekend? Are you going to shop? Then, how about being a savvy consumer? Is Bounty the "quicker picker-upper?" Are expensive shampoos better? Are all teeth whiteners the same, regardless of their prices?

Make questions about all of the commercial advertisements that are so much a part of our lives and become product-testing teams yourselves! :)

Jiyoon

P.S. For more information about consumer product testing, please check the following web-resources!

1. Paper Towel Testing
   http://www.tappi.org/paperu/fun_science/test-Strength.htm
2. Testing Consumer Products
   http://www.scienceproject.com/projects/intro/intermediate/IC004.asp
3. Consumer Testing in Classrooms
   http://www.teachnet.com/lesson/science/scientificmethod/consumertestin g.html

Interdisciplinary Approach

4

How Do You Combine Science with Other Subjects?

I admire Maria in the movie *The Sound of Music* who sings "Do Re Me." She is the role model teacher who already understood the effects of the interdisciplinary approach on learning. Her wonderful integration of music and language art was very educational and effective. In the movie, Maria tried to teach music with this song to the children who did not know about how to sing at all. At the end of the movie, the children could understand easily how to sing and even won the first place at a song competition!

In this chapter, science teachers will learn about the interdisciplinary approach and considerations to develop interdisciplinary curriculum. Through the examples of interdisciplinary activities, teachers will develop ideas of how to integrate their science curriculum with interdisciplinary experiences.

Interdisciplinary Aspects

Science teachers in classrooms need to know how to integrate science with other subjects. Interdisciplinary learning is the natural way that students learn. Students constantly cross disciplines in their daily lives. Teachers integrate into their daily work such areas as psychology, sociology, mathematics, economics, nutrition, safety, communication, drama, music, and the scientific method of discovery. This interdisciplinary approach provides students with the opportunity to confront problems

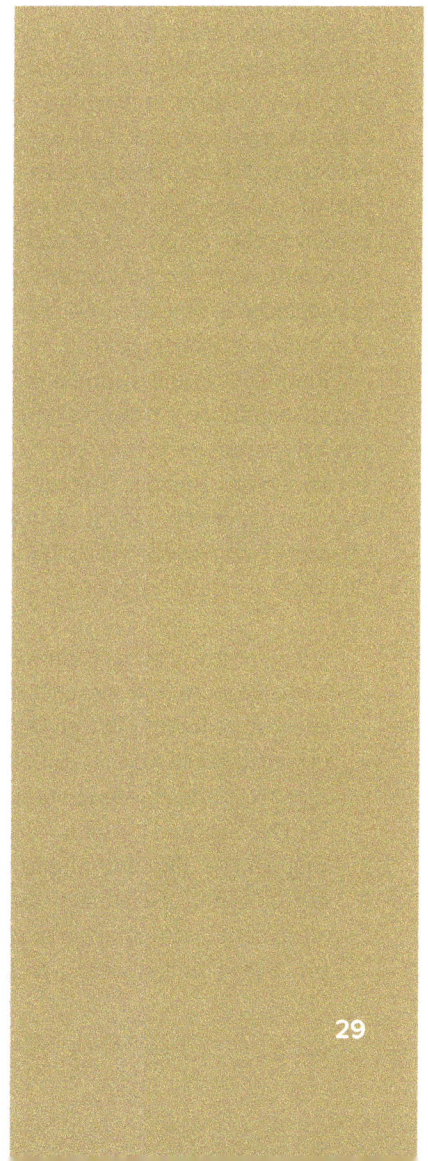

that require multiple and overlapping solutions and to apply their knowledge to real- life situations.

> Much of the curriculum that is contained in textbooks is neither timely nor relevant to students' lives. Additionally, the daily schedule often fragments learning so that each teacher is given a defined time block to cover material that will likely be assessed by a state-mandated test. All of these hindrances make it difficult for teachers to engage students in studying any material in depth and make connections between subject areas and topics. The interdisciplinary model of teaching enables students to see the links between subject areas, [such as] the relationship between literature and history or mathematics and science (Jacobs, 1989).

Previous research studies have demonstrated the effectiveness of science integration with other subject areas on student learning. Hurley (2001) found that integrating science instruction with mathematics was related more strongly than traditional instruction to student achievement in science. Investigations involving the integration of technology have shown to benefit elementary students through high levels of engagement and metacognition (e.g., Amaro-Jimenez, 2008; Jane & Jobling, 1995).

Keith Barton and Lynn Smith (2000) suggest that interdisciplinary learning is especially important in the early grades so as to "provide authentic experiences in more than one content area, offer a range of learning experiences for students, and give students choices in the projects they pursue and the ways they demonstrate their learning."

> [The] interdisciplinary units enable teachers to use classroom time more efficiently and address content in depth, while giving students the opportunity to see the relationship between content areas and engage in authentic tasks.

The National Council for Teachers of English (NCTE) also [emphasizes] the importance of interdisciplinary teaching in a position statement on integration of multiple curricula. Based on discussions from a combined meeting of the major national subject-matter organizations, the NCTE (1995) explains that "educational experiences are more authentic and of greater value to students when the curricula reflect real life, which is multi-faceted—rather than being compartmentalized into neat subject-matter packages." The NCTE highlights the benefits of interdisciplinary teaching and promotes the "natural and logical connections that cut across content areas," which can be organized around "questions, themes, problems, or projects rather than along traditional subject- matter boundaries." ("Curriculum and Instruction," n.d., pp. 3)

Definition of Interdisciplinary Approach

The interdisciplinary approach includes more than one discipline in an area of study, like a sampling of knowledge from different disciplines. The interdisciplinary curriculum is carefully designed, considering scope (range) and sequence (order). Based on the scope and sequence, there are two interdisciplinary models (Martin, 2008):

Daisy Model
The main subject is in the center; other subjects are attached as petals. Many units become a sampling of knowledge from each discipline. "Unlike the disciplines that have an inherent scope and sequence used by curriculum planners, there is no general structure in interdisciplinary work. Curriculum developers must design a content scope and sequence for any interdisciplinary unit or course" (Jacobs, 1989).

Rose Model
The Rose model intertwines all subjects. In the rose, one sees the whole flower regardless of individual petals. Learning focuses on a particular problem or situation that is meaningful and of interest to children, and the children study

whatever necessary to bring about personal understanding without isolating each discipline.

Effective interdisciplinary programs must be carefully conceived of in scope and sequence. Also, teachers must use both discipline field–based and interdisciplinary experiences for students to understand content in the curriculum. Four considerations for developing interdisciplinary teaching are as follows:

- **Discipline-Filled and Interdisciplinary Experience:** "Students should have a range of curriculum experiences that reflect both a disciplinary and an interdisciplinary orientation. Students cannot fully benefit from interdisciplinary studies until they acquire a solid grounding in the various disciplines that the interdisciplinary approach attempts to bridge" (Jacobs & Borland, 1986).
- **Scope and Sequence:** "Teachers must design and implement curriculum based on the scope and sequence of the integrated disciplines. The interdisciplinary units offer students the opportunity to see connections and relevance between topics and provide a variety of perspectives" (Jacobs, 1989). Also, teachers need to be flexible enough to form and revise the curriculum according to the students' needs. Therefore, students should be involved in the planning and development of the interdisciplinary units (Jacobs, 1989).
- **Fragmentation, Relevance, and the Growth of Knowledge:** Interdisciplinary curriculum is used to overcome fragmentation, relevance, and the growth of knowledge (Jacobs, 1989). An interdisciplinary study integrates fragmented insights from reductionist disciplines into holistic understanding. The relevance of knowledge leads to the necessity for interdisciplinary approaches. The remarkable degree of specialization that has resulted from research and practice is found in the field of science. With this blessing and burden of the growth of science knowledge, schools are under pressure to add to science knowledge in

spite of the limited length of school days. Science teachers need to rethink methods for selecting and integrating the various areas of study.
- **Share:** Interdisciplinary units should be shared with all faculty, administration, and community members. By dividing knowledge among specialists, they can contribute their knowledge and skills.

The following steps help teachers to develop interdisciplinary units, including all of the considerations above (Jacobs, 1989):

- Selecting a focus or thematic topic
- Generating ideas or connections between related topics
- Establishing guiding questions for the scope and sequence of the unit
- Designing activities to fulfill the goals of the unit.

Concept Map

To develop interdisciplinary curriculum that considers scope (range) and sequence (order), a concept map is helpful for teachers to see the whole picture of scope and sequence of the curriculum.

A concept map is a map that shows concepts sorted and linked with each other. The concepts are related to each other, but there are ranks among the concepts, depending on how much the concepts cover their sub-concepts. For example, the water concept can be the top concept among evaporation, condensation, and precipitation.

The technique to develop a concept map is

- Create concepts to be learned
- Rank the concepts
- Link among the concepts

The concept map is used to help teachers teach and students understand better science.

Science Activity 4-1 Developing a Concept Map

Activity Procedure:

1. Select a topic for a science unit (for example, the water cycle, plants, solar system...)
2. Develop concepts to teach in the unit (when you choose the water cycle, then the concepts to teach in the unit can be water molecules, evaporation, condensation, precipitation, acid rain...)
3. Rank the concepts
4. Link among the concepts

With the final concept map, you can see how far (scope) and in what order (sequence) you are going to teach.

Strategies for Combining Science with Other Subjects

No subject can be studied in isolation. For example, it is impossible to teach science without some language. Also, science always requires obtaining measurements, calculating data, constructing graphs, and interpreting experimental results that are essential components for mathematics. Social studies provides the essential link between science and its usefulness to society. Therefore, it is necessary to integrate science with other subject areas.

Strategies to be considered when science is integrated with other subjects are as follows:

1. Language arts
   - Using children's reading books dealing with science topics. The reading list recommended by the NSTA is in Appendix E. The example lessons of how to integrate language arts with science can be found in "Picture-perfect language art science lesson."
   - Adding children's writing about science concepts
   - As the classroom activities of how to integrate language art with science, writing in science journals, developing poems about science concepts, and creating foldables can be added to science classrooms.

2. Mathematics
   - Identifying the mathematical core concepts of Numbers/operations, Algebra, Geometry, Measurement, and Data analysis/probability to understand science concepts.
   - Utilizing mathematics tools of: Graphs, Drawings, and Table/chart
   - Connecting Mathematical thinking of: Problem solving, Reasoning/proof, Communication, Connections, Representation, Communication skills, and Communicating results with audience

3. Social studies
   - Bridging science concepts to history and social issues
   - Making decisions through scientific methods about issues of science, technology, and society (STS). For example, the Battle of Changjin Reservoir between 11/26/1950 and 12/13/1950. (12,000 US soldiers could survive even though they fought with 120,000 Chinese soldiers. One of their ways to overcome Chinese soldiers was to build a bridge by analyzing and adapting the geological locations.)

4. Art
   - Increasing student engagement in artful science projects.

5. For example, creating leaf art to describe animals, woods, plants, and so on, developing ornaments of animals, plants, and flowers for a Christmas tree.

Science Activity 4-2 Rock Building

Activity Procedure:

1. Take students out of the classroom and ask them to get together near a pond or a stream where they can find rocks.

2. Ask them to make a rock building by accumulating at least five rocks (you can change the number and the size of rocks depending on the grade of your students). Some of the young children may need help from the teacher.

How it works:

Once the rock building is done, explain to the students that they needed to match the center of the mass of each rock to balance each other. A way of finding the center of mass is as follows:

a. Students hang a rock from a point

b. Mark the vertical line

c. Hang the rock from the second point

d. The center of mass is where the lines cross.

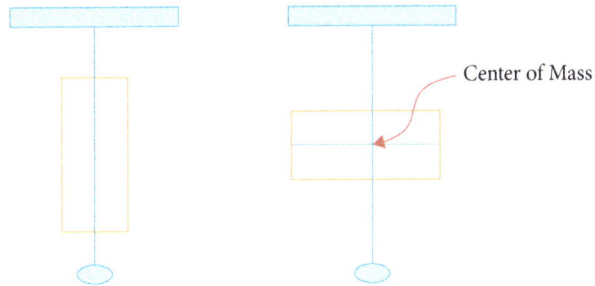

FIGURE 4-1 Finding a center of mass

- Rock building—by finding the center of the mass, students can balance the rocks no matter what they look like (see the Science Activity 4-2).
- Artful learning
  - A new way of learning cycle following the phases of Experience, Inquire, Create, and Reflect
  - Start with a painting; for example, students should observe the painting (experience), make their own experiment (to develop the scene of the painting) (inquire), develop their own painting (create), and reflect on what they found.

5. Music
  - Science Song Contest
    - Developing science songs in groups.
    - Steps:
      - Choose a science concept
      - Develop a concept map about the topic
      - Create lyrics by making sentences using the concepts in the concept map
      - Select a tune from a song that your group is familiar with
      - Develop instruments for the song
      - Present the song to the audience
  - Musical Instrument Contest
    - Sound is the vibration of matter. Students find out how to make sound different by creating various instruments and use the instruments for their science songs.
    - Trash Band—students learn science by creating instruments that are made of trash (recycled materials), thus integrating science with music and a social issue.
  - Musical
    - Creating a musical integrates science with all subject areas, like technology, social studies, music, art, dance, math, and so on.
    - Steps:
      - In a group, develop a story based on the topic of the unit
      - Create a script (Duration is about 10–15 minutes)
      - Add science songs/music
      - Use technology for the stage background and background music and sound

Science, Technology, & Society

The Science-Technology-Society (STS) approach is the most general attempt to overcome challenges in traditional curriculum. In response to a lack of scientific and technological literacy in classroom teaching, various organizations and governmental agencies (American Association for the Advancement of Science 1986; Government of Canada Consultation Paper 1991; National Assessment of Educational Progress 1988; National Science Teachers' Association 1991) have put forward recommendations for curriculum in science education to understand how science and technology shape culture, values, and institutions in societies and how such contexts, in turn, shape science and technology. The National Science Board (NSB, 1983) summarizes the reasons for this emphasis on science and technology in a social context: "Science and technology are integral parts of today's world. Technology, which grows out of scientific discovery, has changed and will continue to change society." STS allows students to learn science and technology in the context of society to develop a better understanding of science, technology, and society altogether.

After the Sputnik Shock in 1957, there have been criticisms on modern science and technology. Educators and scientists have demanded students' abilities to make decisions related to science and society issues and evaluate policies related to science and technology issues, which becomes part of science curriculum for students. However, there has been too much emphasis on college entrance exams and memorization of theories, science concepts, and products.

As the result, textbooks of high schools are revised across the world. In the United States, Project 2061 (American Association for the Advancement of Science (AAAS)) developed CHEMCOM (Chemistry in Community), which integrates chemistry with social issues.

The STS approach affected Korean science curriculum, too. Different from the traditional textbook in Korean schools, the new textbook with the STS approach, called "Next Generation Textbook," includes a social issue at the end of each chapter. For example, at the end of Water Unit, Korean students in 5th grade are asked to develop machines or tools to make the Earth cooler as a solution to the global warming issue.

STS teaching always starts with social issues (invitation) and investigate to find the solutions to the issues (exploration). Once the students reach the solution, they have time to share their ideas (explain) and take action to adapt their solutions to their society and daily lives (see Figure 4-1). Example activities are as follows:

- Acting out roles in a play, writing a letter to policy makers, or drawing a poster
- Water project (exploring the solutions to the water pollution issue)
  - Lesson 1: Introduction of Water Project, National Science Standards, Water Issue, Sea Grant Guest Speaker

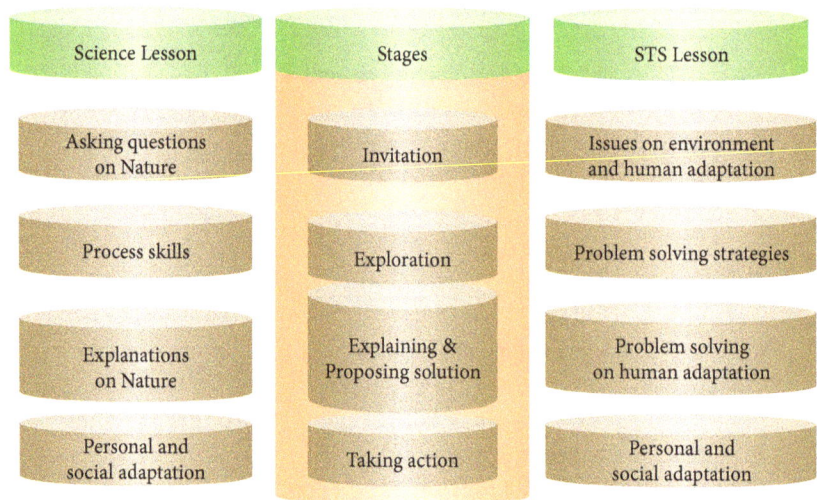

Science Lesson	Stages	STS Lesson
Asking questions on Nature	Invitation	Issues on environment and human adaptation
Process skills	Exploration	Problem solving strategies
Explanations on Nature	Explaining & Proposing solution	Problem solving on human adaptation
Personal and social adaptation	Taking action	Personal and social adaptation

FIGURE 4-2 Learning cycle for STS

Science Activity 4-3 Write a letter to the environmental policy maker

Activity Procedure:
1. Learn about water pollution
2. Discuss solutions to water pollution

3. Divide the class into groups.
4. Explain a situation and ask the groups to write a letter to a policy maker with requirements (see Table 4-1).

TABLE 4-1 **Handout for writing a letter to policy makers for STS.**

Selling water: Handout for "Write a letter to the environmental policy maker"

Situation:

There was a proposal made by a water company for selling Lake Superior water to Asian countries and desert areas where people need water more than regions. NOVA groups and environmentalists argued that allowing water selling would make water levels and flows changed, which will have unpredictable and harmful consequences to basin habitat, biodiversity, shorelines, jobs and culture, particularly to First Nations. Lower water levels will mean greater disturbance of highly contaminated sediments in shallow harbors and connecting channels and less dilution of polluted waters. To this, the entrepreneur of the water company said that draining water would lower the lake an inch, but this would be replenished naturally within 10 hours. He also argued it would be a great chance for the job-poor area.

At 6:00PM tonight, there will be a meeting for this issue at the Community Center. To help him to make a decision, write a letter to the environmental policy maker who is the chair of the meeting.

Requirements for writing a letter:
1. Your group decision (selling or not selling)
2. Reasons for the decision
3. Solutions for the other side
   a. If your group decides not to sell, what will happen to the Asian countries and the desert areas?
   b. If your group decides to sell, what will happen to the Lake Superior habitat?

- Lesson 2: Human relationships to water (biology, aesthetic –cultural–spiritual relationships)
- Lesson 3: Creating posters to educate public people and selecting spaces to post the posters
- Lesson 4: Discussing what solutions to the water pollution issue the students found and how they can apply their findings to other social issues
- Solar-powered classroom
  - Develop a solar panel system that creates electronic power to light the classroom
  - Steps:
    - Bring the social issues related to the natural resources
    - Discuss solutions to save the natural resources (focusing electronic power)
    - Divide the class into groups
    - Work on fundraising online
    - Purchase the solar panel system
    - Install the solar panel system on the roof after school and on weekends together with students and parents
    - Light the classroom

Through this activity, students will gain an understanding of the global water issue and have an opportunity to make a decision about the issue. And, by writing a letter to policy makers, the students learn to educate others.

Community Science

Community Science is one of the instructional methods to integrate science with other subject areas. Our communities offer many opportunities and resources to help children learn science. Students reach outside the classroom to engage community resources in learning science through the community science activity.

Community resources destinations are included in the booklet "Helping Your Child Learn Science" (U.S. Department of Education, 2005):

Zoos

Zoos are great places for science teachers to encourage their students' interest in the natural world and to introduce them to exotic animals that they might not otherwise ever see. Here are a few activities to help make a visit to a zoo worthwhile:

- *Play a guessing game.* Guessing games can help your students understand form and function. For example, ask questions such as the following:
  - Why do you think seals have flippers? (Seals use flippers to swim
  - through the water.)
  - Why do you think these gibbons have such long, strong arms? (Their arms help them swing through the trees.)
  - Why does that armadillo have a head that looks like it's covered with armor?
  - Why is its body covered with those bony plates? (The armor and the bony plates protect it from other animals that want to eat or kill it.)
  - Why is that snake the same brown color as the ground? (As snakes evolved, the brown ones didn't get eaten as quickly.)
- *Match the animals.* Children can learn about organization by seeing related animals. Have them compare the sizes, leg shapes, feet, ears, claws, feathers, or scales of various creatures. The questions teachers can ask the students are:
  - Does the lion look like a regular cat?
  - How are they the same?
  - Does the gorilla look like the baboon?

As the students get older, they will understand more complex answers to these questions.

Museums

In museums, both science teachers and their students can have fun and learn science together. Science and technology museums, natural history museums, and children's museums can be found in many middle-sized and smaller communities, as well as in large cities.

Museums vary in quality. If possible, it is recommended that teachers find museums that have special areas, exhibits, and "hands-on" programs just for children. In these programs, children are often able to use scientific equipment that is far too expensive or specialized for their schools to own. Science teachers should look for museums that have:

- Levers to pull;
- Lights to switch on;
- Buttons to push;
- Animals to pet; and
- Experiments to do.

Planetariums

Planetariums have wonderful exhibits and activities for children. There are over 1,000 planetariums in the United States, ranging from small ones that hold about 20 people to giant facilities with hundreds of seats. These facilities are particularly useful for children who live in urban areas, where city lights and air pollution obstruct the view of the sky. Inside a planetarium, children may be able to:

- Use a telescope to view the rings of Saturn;
- See details of the "sky" from inside the planetarium's dome; and
- Step on scales to learn what they would weigh on the moon or on Mars.

Aquariums

Aquariums enable youngsters to see all kinds of marine life, from starfish to sharks to electric eels, and to learn about their special habitats. Children can enjoy observing feeding. Before visiting an aquarium, it is suggested teachers call ahead to find out when the penguins, sharks, and other creatures get to eat. Also, they should check for special shows that feature sea lions and dolphins.

Farms

A visit to a farm can be a wonderful trip for teachers and their students. Local county extension, farm bureau, or local agriculture offices are places to contact to find a farm to be visited.

- When students visit a dairy farm, teachers encourage the students to ask questions about the cows and their care:
  - What do they eat?
  - Do they sleep?
  - Where is their food kept?
  - What happens to the milk when it leaves the farm?
  - How does it get to the grocery store?
  - Have the students try their hands at milking a cow.
  - How is the equipment used, and how does the milk make its way from the farm to the grocery shelf?

- When students visit a farm that grows crops, teachers encourage them to look at the crops and ask questions about what they see:
  - What crops are grown?
  - How are they planted?
  - How are they harvested?
  - What are they used for?
  - How do they get to the grocery store?
  - If your students grew up in a city, they may have no idea what corn, soybeans, potatoes, or pumpkins look like as they grow in a field.

Science at Work

Students may recognize that many people use science to do their jobs—chemists, doctors, science teachers, computer technicians, and engineers, for example. However, they may not realize that many other jobs also require science skills. To show the students how important science is for many jobs, science teachers arrange for them to spend part of a day—or even an hour—with a park ranger, pharmacist, veterinarian, electrician, plumber, dry cleaner, cook, mechanic, architect, mason, or anyone else whose job involves some kind of science. Before any visit, teachers encourage the students to read about the job so they'll be able to ask good questions. For example, they might ask a dry cleaner questions such as the following:

- What chemicals do you use to clean clothes?
- How are stains removed?
- What happens to the chemicals after you use them?

Community Science Groups and Organizations

Many communities have groups and organizations that include science programs as part of their services for children. Some may sponsor local summer science camps, focusing on areas that range from computers and technology to natural science to space. For example, the Boy Scouts, Girl Scouts, or similar groups; YMCAs and YWCAs; 4-H groups; Audubon; or local colleges and universities.

Other Community Resources

Botanical gardens, weather stations, hospital laboratories, sewage treatment, plants, newspaper plants, recycling centers, and radio and television stations are only a few of the kinds of places in the community where children can learn more about all kinds of science. Other community resources are as follows:

- *Recycling Center or Landfill.* Teachers arrange a tour of a recycling center or landfill to show students what happens to the community's trash. Before the visit, teachers can ask the students to think about questions, such as the following:
  - Where does the trash go when it leaves our homes?
  - What happens to it?
  - How much trash does our community produce each year?
  - What kinds of materials are recycled?
  - What kinds of things can't be recycled?

As science teachers tour the facility, they can ask the students the questions; then, compare the earlier thoughts to what they have learned.

- *Local Water Department or Sewage Treatment Center.* Contact local water department or sewage treatment center to arrange a tour of its facilities. Before the visit, teachers can ask students to think about where the water comes from that they drink and where it goes when it has been used, for example:
  - Is anything added to the water to make it safe to drink?
  - Does all the water used in the community come from the same place?
  - Does all the sewage in the community go to the same place?
  - What happens to the sewage?

The students can compare their earlier answers to what they learn on the tour.

- *Public libraries.* Public libraries are also rich resources for books and magazines on science; videos and DVDs; free Internet access; special programs—such as book talks—that relate to science; and much more.

Through the community science activity, students understand there is useful community information around them, and they can find science in their community and daily lives. As it is presented in natural ways, the students can understand science better.

When teachers plan the community science activity with their students, they need to plan in advance so that there are no time conflicts with other schools that want to visit the locations. When the students go outside the classroom, there must be student activities appropriate to the science concepts that the students can learn at the community science place and their developmental levels that the students have before the community science activity. Because this activity happens outside the classroom, the community science activity needs to be flexible, depending on the weather, the condition of the location, or the workers. When it is raining, then teachers need to prepare activities that can be done inside of the community building or some other activities. Also, teachers should provide another opportunity for the students who cannot participate in the community science activity, so they can learn the science concepts that the students learn from the field trip.

Summary

Interdisciplinary learning is the natural way that students learn. Students constantly cross disciplines in their daily lives. Therefore, science teachers prepare their lessons by integrating daily work, providing them with the opportunity to confront problems that require multiple and overlapping solutions and to apply their knowledge to real-life situations. Teacher can integrate science with language arts, math, social studies, art, music, and other subject areas by using special components for each subject area. STS is one of ways of integrating science with technology and social issues. STS starts with social issues and leads the students to find the solutions to the issues. Example activities of STS approach are writing a letter to policy makers, Water project, and Community Science. The community science activity is another way to integrate science with community resources, like the zoo, library, museum, grocery store, and so on. Integrating with artistic, musical, and artful learning; the science song contest and musical instrument contest are example activities. Effective interdisciplinary curriculum must be carefully conceived for scope and sequence and reflect both discipline field–based and interdisciplinary experiences for students to understand contents. Other considerations to develop the interdisciplinary curriculum are the need to overcome fragmentation, relevance, and the growth of knowledge, working together with faculty, teachers, and members of the community. Also, students need to get engaged in epistemological questions through interdisciplinary learning. With the interdisciplinary approach, teachers can develop a learning environment where the students can understand science in their natural daily lives.

Assignment

Teachers provide science integration with other subject areas. Please choose one of the following subject areas and develop an activity in an interdisciplinary approach. The example activities for each subject will help you to complete this assignment:

- Reading—writing a poem about science concepts, developing foldables
- Math—developing a graph based on data of an experiment
- Social Studies—developing a letter to policy makers, drawing a poster, working on water project or community science activity
- Art—creating leaf art, artful learning, or rock balance activity
- Music—playing a musical instrument, holding a musical instrument invention contest or science song contest

In your activity, include the title of the activity, grade level, and procedure with questions.

References

Barton, K. & Smith, L. (2000). "Themes or Motifs? Aiming for Coherence Through Interdisciplinary Outlines." *The Reading Teacher,* 54(1), 54–63.

Curriculum and Instruction. (n.d.). 3. Retrieved from: https://achieve.lausd.net/cms/lib08/CA01000043/Centricity/Domain/319/PSC%201.0%20Elementary/PROPOSAL.pdf

Jacobs, H. (1989). *Interdisciplinary Curriculum: Design and Implementation.* Alexandria, VA: Association for Supervision and Curriculum Development, 4–5.

Jacobs, H. H. & Borland, J.H. (1986). "The Interdisciplinary Concept Model: Design and Implementation." *Gifted Child Quarterly.* Winter.

Martin, David J. (2008). Elementary Science Methods: A Constructivist Approach, (5th Edition). Belmont, CA: Thomson/Wadsworth Inc., 380–382.

National Council for Teachers of English (1995). "Position Statement on Interdisciplinary Learning, Pre-K to Grade 4." Reviewed http://www.ncte.org/positions/statements/interdisclearnprek4

National Science Board. (1983). *Educating American for the 21st Century.* Washington, DC: U.S. Government Printing Office.

National Science Teachers Association (NSTA). (1991). "The NSTA Position Statement on Science-Technology-Society (STS), *NSTA Handbook,* 47–48.

U.S. Department. (2005). "Helping Your Children Learn Science: With Activities in Children Pre-School through Grade 5." Washington, DC: U.S. Department of Education, Office of Communications and Outreach, 36–46.

Tip of the Day 4

Hi future science teachers,

For your community science, you can think about a place where you have frequently visited. When you have a community science at an aquarium, zoo, planetarium, or museum, you need a time to prepare before you go with your students, for example, contacting professionals, setting up a tour in advance, and getting approvals before visiting the community places. But, by choosing the local grocery store or the outdoor sport store you stop by often, you can save time. You do not have to get any approvals for visiting these places. You already knew about the stores very well and you were the professionals of these places. So, you can do the community science activity right away and most successfully with your students to learn about, for example, seasonal fruits, fishes in the fish tanks, equipment for outdoor activities, or stuffed animals on the wall.

I hope you open your eyes to find many community resources near you.

Jiyoon

5

Technology

How Can You Use Technology in Teaching Science?

'll never forget the 3rd graders' science classroom when I supervised one of the student teachers who practiced teaching in a school for a full semester. In the classroom, there were three computers at the back side of the room, but all of the computers were turned off and nobody used them. I asked one of the 3rd grade students in the room, "Don't you use the computers?" He said "No." But after the "No," I was so shocked that I promised myself that I would not make my teacher candidates hear such a thing from their students. His answer was "my teacher does not know how to use computers. That was the reason why she did not allow us to use the computers." Students in 21st century were born with technology. In many ways, technology is already part of their lives. Especially as a science teacher, it is critical to understand that science goes together with technology. Technology is improved as a product of science and vice versa.

Through this chapter, science teachers will gain definitions, resources, and evaluations of educational technologies. Also, cyberbullying will be discussed as one of the biggest side effects of using technology in education.

Definition of Technology

Everyone has great ideas for using tools to make their lives better and more convenient. For example, to prevent a dish scrubber from being wet all the time, a soap holder can be used as a place where the wet dish scrubber is kept dry while it is not used. Mrs. Robinson, who is a

Science Activity 5-1 Make it Brighter

Materials: cardboard boxes to hold individual materials, D size batteries, wire, #48 bulbs, battery holders, bulb holders, wire strippers, check list, assignment sheet

Activity Procedure:

1. Show students how to make a series circuit with two light bulbs.

2. Provide all the materials to students and ask them to find a way to make the two light bulbs brighter than on the series circuit.

3. When the students find a way to make the two light bulbs brighter, explain that the circuit is the parallel circuit.

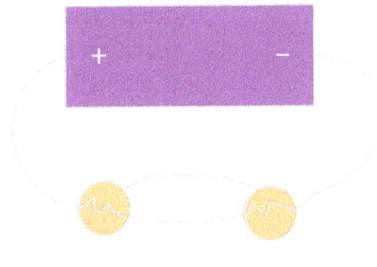

Students understand that there are two kinds of circuits: series and parallel. The parallel circuit makes light bulbs brighter but then use the battery more. On the other hand, the series circuit makes light bulbs dimmer but saves battery.

4. Students have a time to discuss how to use two different electric circuits for their daily lives.

FIGURE 5-1 Series circuit with two light bulbs

FIGURE 5-2 Parallel circuit

dressmaker, has a bag made of zippers. So, when she unzips her bag, then the bag becomes a long roller to measure. These kinds of great ideas can be called by technology. Also, technology can be material products, like computers, cameras, etc. There are four different meanings in technology:

- Technics: application of mental and physical effort in order to achieve some value, tools and machines that may be used to solve real-world problems (watch, knife, videocassette,...)
- A technology: scientific method of achieving a practical purpose (carburetor technology, brake technology, transmission technology,...)
- A form of human cultural activity: activity that forms and changes culture (art, law, sport, religion,...)
- A total societal enterprise: an organization that applies strategies to maximize improvements in human and environmental wellbeing)

Not only the single material products of science, but also methods and human cultural activities are called by technologies. When teachers talk about using technology in science classrooms, it is easy for them to think about only computers. But the true definition of technology means the methods and ideas of how to use the products according to their needs. It is important for science teachers not only to know what kind of educational software and hardware are out there but also to understand how to use them in their science classrooms effectively, according to students' needs.

This concept can be compared to the Piano Stairs in Odenplan, Stockholm. Many people know about pianos and escalators separately, but nobody thinks about combining pianos with escalators. Scientists in Stockholm worry about people's health. People do not exercise at all. Most of the people like to use escalators instead of stairs. So, the scientists thought about constructing the Piano Stairs that makes piano sound when people step on stairs. As a result, 66% of the escalator users walked on the piano stairs instead of easy riding on the escalator. When people know about the relationship between science and technology and adapt it to human society, science and technology work better and become useful for human life.

Integration of Educational Technology in Science Classroom

When science teachers develop an interdisciplinary lesson, the Frayer Model is another great way to integrate science with language art. It is a graphic organizer for building science vocabulary. This model helps students better understand complicated science concepts by identifying and comparing their definitions, characteristics, examples, and non-examples. Figure 5-3 shows the diagram of the Frayer model.

The additional activity with Wordle after the Frayer Model enhances students' understanding of the concept. The Wordle is a group of words that are gathered in different font colors, sizes, and arrangements. This format is useful for quickly grasping the most outstanding words and placing words alphabetically. Also, Wordle has a function that can distinguish words by their meaning, rather than appearance. Further,

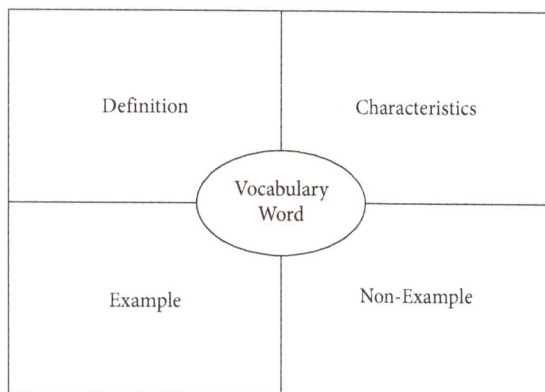

FIGURE 5-3 Frayer Model

the word terms can hyperlink to the items associated with the words. Once the words that are understood by the students through the Frayer Model and they are organized in Wordle, the words are meta-cognitively imprinted in their minds with the different organization and linked explanations.

Technology cannot replace education. However, when technology is included in education, the effects of education can be expelled and boost the next level of performance. Without technology, education can happen, but with technology, education can be accomplished better. It is recommended that teachers include educational technology in their teaching science.

Web Tools

Innovative science teachers frequently use intuitive programs and websites that are beneficial for teaching and learning science. The web tools can save the teachers' time and efforts and help the students to understand science better by sharing and continuing their learning before, during, and even after the class. The web tools that are recommended to use in classrooms are:

- Social Networking
  Delicious (http://delicious.com/)
  Diigo (http://www.diigo.com/)
  Edmodo (http://edmodo.com/)
  Google Business (Google Docs, chat, forms, and so on)
  (http://www.google.com/intx/en_sg/work/apps/business/products/)
  Personalized Home Pages (http://www.google.com/ig,
  http://www.pageflakes.com/, or http://www.netvibes.com/)
  Twitter (http://www.twitter.com/)
  WordPress (http://wordpress.org/)
  EduBlogs (http://edublogs.org/)
- Digital Images
  Flickr (http://flickr.com/)

Flickr Storm (http://www.zoo-m.com/flickr-storm/)
Flickr Toys (http://bignugelabs.com/flickr/)
Spell with Flickr (http://metaatem.net/words/)
More Flickr Toys (http://www.pimpampum.net/toys/)
Visualizeus (http://vi.sualize.us/)

- Organizers
Awesome Highlighter (http://www.awesomehighlighter.com/)
Chart All (http://www.chartall.com/)
Doodle (http://www.doodle.com/)
Evernote (http://evernote.com/)
Dropbox (http://dropbox.com/)
Glogsgter (http://www.glogster.com/)
Studiyo (http://mystudiyo.com/)
Tiny URL (http://tiny.cc/) & (http://tinyurl.com/)
Wordle (http://www.wordle.net/)
Gliffy (http://www.gliffy.com/)
Bubbl.us (http://bubbl.us/)
xTimeline (http://www.xtimeline.com/)

- Open Education Resources
OER Commons (http://oercommons.org/)
EduTopia (http://edutopia.org/)
Harvard Graduate School of Education (http://gse.harvard.edu/)
Teaching Channel (http://teachingchannel.org/)
Khan Academy (http://khanacademy.org/)
EdX (http://edx.org/)
MIT Open course ware (http://ocw.mit.edu/)
Ted Ed (http://ed.ted.com/)

iPad Applications

Using iPads in classrooms is an emerging trend across the world. Everyone likes to use the touch screen for more interaction and motivation in learning. Korea is one of the countries where students use iPads popularly in school curriculum. Korean schools provide not only iPad hardware, but also digital curriculum software downloadable on the iPads, in contrast with schools in countries with limited availability of iPad curriculum software.

There are many advantages of using iPads in classrooms:

- Ease of use
- High motivation and engagement
- No need for heavy backpacks
- Open resources
- Readable foreign resources
- Better communication among students, teachers, and parents
- Self-directed learning

In contrast to these advantages, science teachers need to prepare themselves for disadvantages:

- Shortage of teacher readiness
- Limited availability of school curriculum
- Expensive to purchase
- Easy to break
- Complicated after-care service

One of the iPad applications mostly used in science classrooms is Interactive Open Educational Resources (OERs). Numerous researchers recognized the importance of using *interactive OERs* in science classrooms for students' success (Chiong, et. al., 2012; DeLoache & Chiong, 2010; Risconscente, 2012). The OERs excel and lead the transition from passive to *interactive* learning by *using* digital resources instead of traditional physical books. Especially as the *interactive* OERs that include interactive lessons and simulations are emerging as a powerful tool that transforms online and offline teaching and learning, American classrooms are increasing their use of interactive OERs with students. Thus, it is necessary to prepare teacher candidates with this powerful teaching tool that can help their students become actively involved in learning online and offline (Yoon, 2017).

There are so many *interactive* OERs available online that teacher candidates should not have difficulty finding these resources, but it is a challenge to judge their quality and relevance for teaching (Yoon, 2017). A rubric for exploring and rating interactive OERs with is provided in Appendix F. With the interactive OERs that are effective for teaching science, the teachers facilitate an environment where teachers and students are interactive.

As one of the important components in evaluating the interactive OERs, previous research studies consider social rapport and instructional design. Social rapport and increased collaboration can lead to greater levels of interaction that address instructional goals. Another component to be included in well-researched criteria for evaluating the interactive OERs is instructional design. Interactive OERs need to encourage reflection and discussion on the topics and concepts to be learned. The instructional designs of the interactive OER serve to increase this kind of participation and feedback from the learners (Kimeldorf, 1995; Roblyer & Ekhaml, 1999; Yoon, 2017).

In assessing the interactive qualities of interactive OERs, the impact on students' learning is important to count in the constructional approach. To measure the interactive qualities of interactive OERs, it is necessary to see how much students construct their learning through teaching with the interactive OERs (Wilson & Peterson, 2006). McHenry and Bozik (1997) point out that students respond to effectively (or ineffectively) designed interactive OERs with observable behaviors. This dimension "evidence[s] itself most often in an increased or decreased willingness to use the interactive OERs to collaborate with other students, to take responsibility for requesting needed information from the instructor, and to participate in class activities" (Yoon, 2017).

Video Games

"Take two aspirin and call me in the morning." This is what a doctor said as a prescription to a patient before. But now, the prescription can be changed to "play two games and call me in the morning." Video games can replace aspirin pills. There is no pain while people play with video games. What this means is that the video games can be motivational and educational, depending on how you approach them.

These days, young children can easily be found to have handheld video games. It is a big trend that children spend most of their free time playing video games.

Rather than forcing students to not play video games, wise science teachers see the values of video games for doing science. Students can learn science from video games. Teachers can ask their students to investigate how video games are made, like game design, graphics, audio, and game programming. Also, students can explore the usefulness and impact of video games, for example, human behavior during game play, the effects of social or educational content in games, the usefulness of simulations in training and experimentation, and ergonomics.

Cyberbullying

Cyberbullying is the newest form of bullying in schools. It has emerged as students have become more adept at using computers, cell phones, other mobile devices, as well as Internet applications for communication, education, and socialization. Topics of abuse can be the same in cyberspace as in face-to-face communication. However, because of the rapidity and non-physical presence between victim and perpetrator, cyberbullying can become more dangerous to students in schools (Levy, 2011). Cyberbullying is any harassment that occurs via the Internet, computers, and/or mobile technologies and their applications. Cruel forum posts, posting fake profiles on websites, and mean email messages are examples of cyberbullying.

Serious physical threats and suicides have resulted from cyberbullying. This grave problem was first reported among teenagers (Englander, Mills & McCoy, 2009) and now cyberbullying affects children under 13 years old (Bauchner, 2011). According to the National Institutes of Health (2010), "the prevalence of bullying is high, with 20.8 percent of U.S. adolescents in school having been bullied physically at least once in the last two months, 53.6 percent having been bullied verbally, and 51.4 percent bullied socially (excluded or ostracized), and 13.6 percent having been bullied electronically." Also "95% of social media-using teens who have witnessed cruel behavior on the sites say they have seen others ignoring the mean behavior; 55% witness this frequently" (Lenhart et al., 2011, p. 20).

Laird (2012) found that 7.5 million Facebook users are under 13 years old and 20% of children cyberbullied think about suicide and 1 in 10 attempt it.

The negative effects on children and adolescents have been evidenced in a variety ways (Ang & Goh, 2010; Bauchner, 2011; Beale & Hall, 2007; Englander, Mills & McCoy, 2009). Cyberbullying can be a very lonely problem, as students may be home alone when it occurs, and they may not report it to their parents, so the abuse may continue for extended periods (Juvonen & Gross, 2008). Additionally, according to Price and Dalgleish (2010), cyberbullying negatively impacts students in many ways: self-esteem and confidence, plus relationships and scholarly activity, can deteriorate with an increase in anger and depression, as well as self-harming and suicidal thoughts.

According to the National Institutes of Health (2010), bully-victims are more likely to report feelings of depression than are other groups, which interferes with scholastic achievement, social skills, and feelings of well-being. This can, in turn, cause an obstacle to students in schools.

To protect students from bullying, the U.S. Department of Education had released a comprehensive review of state anti-bullying laws and policies encompassing the discussion of school district policy development (U.S. Department of Education, 2011). The report noted that, at the time, 36 states had language of anti-cyberbullying in laws or policies. Thirteen states extended a school's jurisdiction over cyberbully cases to both on- and off-campus and/or through the school's technological infrastructure if the behavior created a hostile campus environment. "Cyberbullying and related behaviors are addressed in a single or in multiple laws. In some cases, cyberbullying appears in the criminal code of a state that may apply to juveniles" (U.S. Department of Education, 2011).

As technology becomes more ingrained in education, the negative effects of cyberbullying have to be acknowledged. It interferes with the promising future of young children. Many students in schools are experiencing cyberbullying by their peers, but there are few research studies about how to protect them from the cyberbullying attacks. Fortunately, the U.S. Department of Education (2011) has taken cyberbullying seriously by reviewing state anti-bullying laws and model policies.

Evaluation of Educational Technology

To choose the educational technology appropriate to specific science lessons, science teachers need to evaluate educational technology in a pedagogical and technical dimension and based on International Society for Technology in Education (ISTE) Standards.

1. *Pedagogical dimension.* The pedagogical dimension has two assumptions in assessing educational technology: constructivism and behavior assumption.

When educational technology is evaluated on constructivism assumptions,

- Learning by using the educational technology needs to be an active and individualized process.
- The learner who uses the educational technology must construct new knowledge based on his/her own individualized experience.
- The learner is the producer of information rather than the consumer.
- The teacher is a member of learning community rather than the only source of information (facilitator).
- Learning emphasizes the application of knowledge in real-life situations.
- The educational technology handles multi-intelligence and assesses learners.

In behavior assumption,

- The educational technology is controlled by the designer (teacher-directed approach).
- Learning by using the educational technology is manipulated by the teacher.
- Learning is described as a stimulus-and-response relationship.
- The student can't study a new topic before completing the prior topic.

- Learning processes can be studied most objectively when the focus of study is on stimulus and responses.
- Drill and practice software are good examples of educational technology in behavior assumption.

2. *Technical dimension.* In the technical dimension, educational technology needs to be evaluated by considering the following aspects:

- Navigation
- Presentation
- Technical information
- Documentation, like a teacher's guidebook or directions

International Society for Technology in Education Standards for Teachers (ISTE, 2008)

Effective science teachers model and apply the International Society for Technology in Education (ISTE) Standards for Students (Standards-S) as they design, implement, and assess learning experiences to engage students and improve learning; enrich professional practice; and provide positive models for students, colleagues, and the community. All teachers should meet the following standards and performance indicators:

- Facilitate and inspire student learning and creativity
- Design and develop digital-age learning experiences and assessments
- Model digital age work and learning
- Promote and model digital citizenship and responsibility
- Engage in professional growth and leadership

Therefore, the educational technology that the teachers use need to follow the ISTE standards for teachers promoting student learning.

Summary

Technology is defined in four different ways: technics, a technology, a total human cultural activity, and a total societal enterprise. When teachers want to use technology in classrooms, the technology means not only a computer (hardware) but also how to use it (software/method). For the better usage of technology, teachers need to know what kinds of technology there are and how the technology can be used for classrooms. There are so many emerging web tools and iPad applications that can be used in schools, but innovative science teachers need to know how to choose the technology appropriate to teaching their students in a pedagogical and technical dimension. Also, science teachers are required to make students aware of the serious side effects of using technology, like cyberbullying.

Science Activity 5-2 Teaching with Interactive Open Educational Resources

Activity Procedure:

1. Use the sample interactive Open Educational Resources (OERs) that are provided in Appendix F to integrate your inquiry-based Learning Cycle lesson with (at least one) interactive OERs.

2. You can select your own interactive OERs based on the rubric provided in Table 1 of Appendix F that is given to use as criteria in selecting the interactive OERs to teach.

Table 5-1 presents the sample interactive OERs that can be used for developing science lessons according to content. Table 5-2 is the rubric modified from the model of Roblyer and Ekhaml (2000). Use the keywords that are shown in the table to find the interactive OER sites.

TABLE 5-1 Sample interactive OERs to teach science. (Yoon, 2017)

Contents	Physical Science	Life Science	Earth/Space Science
Interactive OERs	• 101 in 1 Physics Solver • Active Sonar • Alchemy Glossary • AP Physics • Atom in a Box • Chemical Equation • Chemical Formulas • Chemistry Formulas • Chemistry Terms • Colour Collider • Dictionary of Chemistry • Dictionary of Physics • Elemental Table Formulary: Physics • Gear Ratio • iChemistryLab • iLab Timer • Molecules • Mr Science Show • Mythbusters • Newton's Cradle Physics • Newton's Laws • Oxford Dictionary of Chemistry • Pendulums • Periodic • Periodic Table • Physics Formulas • Physics Puzzles • Physics XL • Physiology Glossary • Wolfram General Chemistry Course • XChem • Touch Physics • Toy Physics • Titration Simulator • Rocket Universe	• 3D Brain • 3D Cell Simulator • 3D4Medical • A Life Cycle App • Anatomy and Physiology • Anatomy Flash • Bio Dictionary • Biology Memory • Biology Body Parts - Human Body • Reproduction • Bugs and Insects • Bugism • Buzz Aldrin Portal to Science • Cellular Biology Digestion • Frog dissection • Genetic Decoder • Genetic History • Genetics Study Guide • HD Marine Life • Human Anatomy • iAnatomy • Insects HD • Marine Life • Nanosaur 2 • Nature Human Genome • Rat Dissection • Respiratory System • Virtual Frog Dissection	• 8 Planets • Astronomy HD • Beautiful Planet • Cosmic Discoveries • Deep Sky • Earth Observer • ExoplanetGravity Balls • Google Earth • Grand Tour 3D: Pocket Solar System • HD Astronomy • HD Solar System • Jupiter Study Guide • Mars Globe HD • Mars Study Guide • Moon Globe HD • NASA • Orion StarSeek PRO • Planets • Planet's New • Pluto Study Guide • WeatherBug • Venus Study Guide • Saturn Study Guide • Solar System • Solar System Guide • Space Images • SpaceTime for iPad • Star Chart • Star Gazer • StarMap 3D Plus • Star New • Stars • Star Walk • Stellarium XL • The Weather Channel • Solar System Simulation

The rubric shown below in Table 5-2 has been modified from the model of Roblyer and Ekhaml (2000) and the Constructivism Approach (Wilson & Peterson, 2006) and has three separate dimensions that contribute to a level of interaction/interactivity and construction of knowledge.

RUBRIC DIRECTIONS: The rubric shown below has three separate elements that contribute to a level of interaction and interactivity. For each of these three elements, circle a description below it that applies best to the interactive OER that you find online. After reviewing all elements and circling the appropriate level, add up the points to determine the level of interactive qualities (e.g., low, moderate, or high). The site that has the highest score is the best site for your lesson.

- Low Interactive qualities, 1–7 points
- Moderate interactive qualities, 8–14 points
- High interactive qualities, 15–20 points

TABLE 5-2 **Rubric for assessing interactive qualities of open educational resources. (Adapted from Yoon, 2017)**

Scale (point)	Element #1 Social Rapport-building Activities	Element #2 Instructional Designs for Learning	Element #3 Levels of Constructivism Approach
Few interactive qualities	The interactive OERs do not encourage students to get to know one another on a personal basis. No activities require social interaction	The interactive OERs do not require two-way interaction between instructor and students; they call for one-way delivery of information (e. g., instructor lectures, text delivery)	The interactive OERs do not encourage students to construct new knowledge based on their own individualized experience, Not emphasizing the application of knowledge in real life situations and caring multi-intelligence
Minimum interactive qualities (2 points each)	The interactive OERs provide for exchanging personal information among students	The interactive OERs require students to communicate with the instructor on an individual basis only	The interactive OERs encourage students to construct new knowledge based on their own individualized experience, but not emphasizing the application of knowledge in real life situations and caring multi-intelligence
Moderate interactive qualities (3 points each)	The interactive OERs provide more than one activity designed to increase social rapport among students	In addition to the requiring students to communicate with the instructor, the interactive OERs require students to work with one another	The interactive OERs enable students to construct new knowledge based on their own individualized experience, emphasizing the application of knowledge in real life situations but not caring multi-intelligence

Above average interactive qualities (4 points each)	*The* interactive OERs provide several activities designed to increase social rapport among students	In addition to the requiring students to communicate with the instructor, the interactive OERs require students to work with one another (e. g., in pairs or small groups) and share results with one another and the rest of the class	The interactive OERs enable students to construct new knowledge based on their own individualized experience, providing a few opportunities
High level of interactive qualities (5 points each)	In addition to providing for exchanges of personal information among students, the interactive OERs provide a variety of activities designed to increase social rapport among students	In addition to the requiring students to communicate with the instructor, the interactive OERs require students to work with one another (e. g., in pairs or small groups) and outside experts and share results with one another and the rest of the class	The interactive OERs enable students to construct new knowledge based on their own individualized experience, emphasizing the application of knowledge in real life situations and caring multi-intelligence
Total for each	_____ pts.	_____ pts.	_____ pts.
Total overall:	_____ pts.	_____ pts.	_____ pts.

References

Ang, R.P., & Goh, D.H. (2010). Cyberbullying among adolescents: The role of affective and cognitive empathy, and gender. *Child Psychiatry and Human Development, 41*(4), 387–397.

Barton, K. & Smith, L. (2000). "Themes or Motifs? Aiming for Coherence Through Interdisciplinary Outlines." *The Reading Teacher.* 54(1), 54–63.

Bauchner, H. (2011, April 20). "Benefits and risks of social media use in children and adolescents." *Journal Watch Pediatrics & Adolescent Medicine.*

Beale, A.V., & Hall, K.R. (2007). "Cyberbullying: What school administrators (and parents) can do." *Clearing House: A Journal of Educational Strategies, Issues and Ideas* 81(1), 8–12.

Chiong, C., Ree, J., Takeuchi, L., & Erickson, I. (2012). "Print Books vs. E-books: Comparing parent-child co-reading on print, basic, and enhanced e-book platforms." The Joan Ganz Cooney Center at Sesame Workshop, New York.

Chmiliar, L. (2013). "The iPad and Preschool Children. Journal on Technology and Persons with Disabilities," Retrieved March 31, 2015 from http://lmcacademy.com/training/amazing-research-results-the-ipad-and- preschool-children-with-learning-challenges

DeLoache, J. S. & Chiong, C. (2010). "Babies and Baby Media." *American Behavioral Scientist*, 52, 1115–1137.

Englander, E., Mills, E., & McCoy, M. (2009). "Cyberbullying and information exposure: User-generated content in post-secondary education." *International Journal of Contemporary Sociology, 46*(2), 213–230. Retrieved March 31, 2015 from http://webhost.bridgew.edu/marc/user-generated%20data%20englander%20mills%20mccoy.pdf

Griffith, M. (2002). Educational Benefits of Videogames. *Education and Health, 20* (3), 47–51.

International Society for Technology in Education. (2008). ISTE Technology Standards. Retrieved from: https://www.iste.org/docs/pdfs/20-14_ISTE_Standards-T_PDF.pdf

Jacobs, H. (1989). *Interdisciplinary Curriculum: Design and Implementation.* Alexandria, VA: Association for Supervision and Curriculum Development. 4–5.

Jacobs, H. H. & Borland, J.H. (1986). "The Interdisciplinary Concept Model: Design and Implementation." *Gifted Child Quarterly.* Winter.

Juvonen, J. & Gross, S. (2008). "Extending the School Grounds? Bullying Experiences in Cyberspace." *Journal of School Health,* 78(9), 496–505.

Laird, S. (2012). "Cyberbullying: Scourge of the Internet", Mashable, Retrieved from http://mashable.com/2012/07/08/cyberbullying-infographic/#7IA_b41yUZqc

Levy, P. (2011). "Confronting cyberbullying: Experts say that schools need to stop worrying about external internet predators and take on the threat within: Cyberbullying." *The Journal Technological Horizons In Education, 38*(5), 25.

Lenhart, A., Madden, M., Smith, A., Purcell, K., Zickuhr, K., Rainie, L. (2011, November 9). "Teens, kindness and cruelty on social network sites." Pew Internet & American Life Project. Retrieved from http://pewinternet.org/Reports/2011/Teens-and-social-media/Summary.aspx?view=all

Martin, David J. (2008). *Elementary Science Methods: A Constructivist Approach,* (5th Edition). Belmont, CA: Thomson/Wadsworth Inc., 380–382.

National Council for Teachers of English (1995). "Position Statement on Interdisciplinary Learning, Pre-K to Grade 4." Reviewed http://www.ncte.org/positions/ statements/interdisclearnprek4

National Institutes of Health (2010). *Depression high among youth victims of school cyber bullying, NIH researchers report. Retrieved March 27, 2013 from http://www.nih.gov/news/health/sep2010/nichd-21.htm*

Price, M., & Dalgleish, J. (2010). "Cyberbullying: Experiences, impacts and coping strategies as described by Australian young people." Youth Studies Australia, 29, 51–59.

Risconscente, M. (2012). Mobile Learning Game Improves 5th Graders' Fractions Knowledge and Attitudes. Los Angeles, CA: GameDesk Institute, Retrieved March 31, 2015 from http://

gamedesk.org/project/motion-math-in-class/

Roblyer, M. D., & Ekhaml, L. (2000). "How interactive are your distance courses? A Rubric for Assessing Interaction in Distance Learning." Online Journal of Distance Learning Administration [On-line serial], 3(2). Retrieved March 31, 2015 from http://www.westga.edu/~distance/roblyer32.htm

Karsenti, T. & Fievez, A. (2013). "The iPad in education: Uses, benefits & challenges, Preliminary Report of key findings," Retrieved May 17, 2015 from http://karsenti.ca/ipad/iPad_report_Karsenti-Fievez_EN.pdf

Troutner, J. (2002). "Best Web 20 Tools, Creative Computer Enterprise," Retrieved May 17, 2015 from http://www.scribd.com/doc/37440092/Best-Web-20-Tools

U.S. Department of Education. (2011). "Analysis of State Bullying Laws and Policies." Retrieved March 26, 2013 from http://www2.ed.gov/rschstat/eval/bullying/ state-bullying-laws/state-bullying-laws.pdf

Wilson, S. & Peterson, P. (2006). "Theories of Learning and Teaching: What Do They Mean for Educators?" (Working paper). Washington, DC: National Education Association Research Department.

Yoon, J. (2017). "Developing a List and Rubric of Interactive Open Education Resources (OER) for Science Teacher Candidates of Diverse Students." *TEM Journal, 6*(3). 512–524.

Tip of the Day 5

Hi future science teachers,

I am not good at parking. It is more difficult when I need to park in my small garage. To resolve this issue, I made a ball called the "Sweet Spot" (see the orange ball hanging on ceiling in Figure 5-4). This ball guides me to the spot where I need to stop my car inside the garage. Since having this ball, I have not had to get out several times from my car to see if I parked in the right spot and have not had any accidents caused by pulling my car forward too much. I love this idea. I am so proud of myself and my technique (using tools and applying them to life appropriately).

Jiyoon

FIGURE 5-4 Sweet spot

Diversity

How Can You Take Care of Diverse Students?

6

After Pluto passed away, I adopted a Yorkie/Pomeranian mix named Mickey. Mickey is a very active and cute little dog (see Figure 6-1). Whenever I invited American friends, they always admired how cute he was. Mickey welcomed them by strongly shaking his tail and making a happy barking. Contrary to those visitors, when my Asian friends visited me, at least one of them said "What a meal for one!" as soon as he was welcomed by Mickey. Some of you might have a cultural shock because of the different reaction of my friends to Mickey. But to me, they are all my friends. I can accept them because I understand both American and Asian friends. I know their different cultures.

When science teachers go to science classrooms, the classrooms are full of diverse students with different cultures, families, and learning styles. Science is for all. Every one of your students in science classrooms has a right to learn science and science teachers are required to teach them using different teaching styles to meet the diverse needs of the students. Before using the different teaching styles for diverse students, teachers need to have acceptance regardless of judgement or fear of difference. Throughout this chapter, science teachers understand different instructional methods to take care of diverse students in the science classrooms.

FIGURE 6-1 Mickey

Piaget's Developmental Stages

According to Jean Piaget, children progress through a series of four key stages of cognitive development. Each stage is marked by shifts in how children understand the world. Piaget believed that children are like little scientists and that they actively try to explore and make sense of the world around them. Through his observations of his own children, Piaget developed a stage theory of intellectual development that included four distinct stages. (Moore, 2014)

- the sensorimotor stage, the children from birth to age 2 begin to interact with the environment;
- the preoperational stage, children from age 2 to about age 7 begin to represent the world symbolically;
- the concrete operational stage, children from age 7 to 11 learn rules, such as conservation;
- and the formal operational stage, adolescents to adults can transcend the concrete situations and think about the future.

When a science teacher enters an elementary classroom, he or she may see many students in the concrete operation stage who can understand rules. But, the teacher also needs to understand that there might be students in either the preoperational or formal operation stage. Also, it is possible that there are students in the sensorimotor stage. Therefore, science teachers prepare their lessons for taking care of all of the students who are in different developmental levels.

Intellectual growth involves three fundamental processes: assimilation, accommodation, and equilibration. Assimilation involves the incorporation of new events into preexisting cognitive structures. Accommodation means existing structures change to accommodate to the new information. This dual process, assimilation- accommodation, enables the child to form schema. Equilibration involves the person striking a balance between himself [or herself] and the environment, between assimilation and accommodation. When a child experiences a new event, disequilibrium sets in until he [or she] is able to assimilate and accommodate the new information and thus attain equilibrium. There are many types of equilibrium between assimilation and accommodation that vary with the [developmental] levels of [students] (Lavatelli, 1973).

All the students in science classrooms have the same fundamental processes for the intellectual growth, but they have various intelligence content and structure depending on their backgrounds. For example, when the students have different families and cultural backgrounds, their content of knowledge and their way of constructing knowledge are variable. Before they go into the classroom, science teachers should understand that the students in the science classroom have multiple intelligences, like nature smart (naturalist), people smart (interpersonal), number smart (logical/ mathematical), picture smart (visual/spatial), self-smart (intrapersonal), body smart (bodily/kinesthetic), music smart (musical), and word smart (linguistic).

Globalized Science Curriculum

Globalization of the classroom is a popular issue these days. From 2003 to 2004, there were about five million Limited English Proficient (LEP) students in public schools, both foreign-born and U.S.-born, residing in the United States. This number represents about 10.1 percent of total public-school student enrollment in the United States. The total number of LEP students has grown by 65 percent since 1994 and have established a relatively large presence in California, Texas, and New York (Batalova, 2006).

For the diverse cultural backgrounds of students in science classrooms, teachers prepare their lessons in globalized curriculum, incorporating international

Science Activity 6-1 Heavy Newspaper with Tae Kwon Do (Korean Martial Art) (Spangler, 2013b)

Materials: 8 inches of ruler (with one inch wide), One Large sheet of newspaper, A Table.

Activity Procedure:
1. Place the ruler on a table and let one end hang over the edge about 4 inches.
2. Use a single sheet of newspaper to cover the portion of the ruler that is lying on the table. Make sure that the newspaper is flush with the edge of the table.
3. Ask the students, "What do you think will happen now if I hit the ruler with the unfolded newspaper covering the ruler?" You might anticipate an answer like, "The newspaper will go flying … or the sheet of newspaper will tear apart."
4. Smooth down the newspaper with your hands so that there are no pockets of air under the sheet of paper. Make sure everyone is out of harm's way as you hit the ruler.
5. Strike the edge of the ruler with a hand strike of Taekwon-do (a sudden sharp hit). The ruler breaks.

How it works

The results of the experiment prove that the newspaper is more difficult to lift when it is spread out over a large area, yet the weight of the folded and flat newspaper remains the same. What other force is exerted on the newspaper that could account for these differences? The answer is the pressure of the air pushing downward on the newspaper that prevents the paper from rising.

It might be useful to picture a giant column of air resting on top of the newspaper. This column of air is 80 miles tall! This column of air above the newspaper pushes down with a force of 14.7 pounds of pressure per square inch (this is at sea level). In other words, each square inch of the newspaper has 14.7 pounds pushing down on it. So, if you know the area of the newspaper, you can calculate the total amount of pressure pushing downward on the paper. Let's say that the newspaper dimensions measure 20 inches by 30 inches. The total area is 20 inches x 30 inches = 600 square inches. If each square inch has a force of 14.7 pounds pushing on it, then 600 square inches x 14.7 pounds per square inch = 8,820 pounds! That's the equivalent weight of two large automobiles. It's no wonder that the newspaper stayed in place at the moment when you hit the stick with a fast strike, which was possible with Taekwondo (Yoosma, 2012). Smoothing down the newspaper with your hands prior to hitting the stick is also a crucially important step. You want to make certain that there is no air under the newspaper that might help it to lift up when you strike the stick.

This activity can be combined with any other martial art that has a fast hand strike to include various cultures in teaching science

elements in existing courses, using examples, case studies, and guest lectures, creating modules based on international themes as part of the course, and integrating an international field trip (short-term international travel). Examples of globalized science curriculum combined with Korean culture are provided in the following:

Science Activity 6-2 Flat Stanley Project to Change the World Weather (Hubert, 2011)

Materials:

- A paper "Flat Stanley"
- Any places on a specific day
- A camera
- The Internet

Activity Procedure:

1. To begin the Flat Stanley Project, first register and log in on the Homepage of The Flat Stanley Project.
2. Next, check out the Flat Stanley List of Participants to see where you might like to send Flat Stanley; who might be sending a Flat Stanley to you; or what classroom, and in what country you would like to arrange an exchange . (Please note, you must have registered and be logged in to view the List of Participants.) It's a good idea to contact your recipient by email at this stage to confirm the exchange.
3. Once you have coordinated a recipient of your Flat Stanleys or flat characters, have your students make a Flat Stanley. This can be done any way your students want to, but most look at pictures of other Stanleys, roughly stencil a new Stanley shape, and color and accessorize as your students see fit. When your Flat Stanley is ready for his journey, be sure to include the sender's name, return address, and email on the back. You might want to document Stanley's starting point in your classroom or home, with a picture, journal entry, or biography about your own flat character.
4. Take a picture with Flat Stanley with specific weather. For example, in March of Minnesota, the students can take a picture of Stanley with Snow.
5. Either mail or email Flat Stanley on his or her journey to the area around the world where the weather will be different, for example to Dallas, TX, or South Africa. All sorts of good teaching activities can be involved at this stage: geography, with locations of Stanley's travels and destinations; math, in distances and times; narrative and writing, with journal entries and biographies; and on and on. Track Stanley's journey as he makes his way to your chosen destination.

How it works

The basic principle of The Flat Stanley Project is to connect students or classrooms with other children or classrooms participating in the project by sending out "flat" visitors, created by the children, through the mail (or digitally, with The Flat Stanley app). Students then talk about, track, and write about their flat character's journey and adventures. Although similar to a pen pal activity, Flat Stanley is actually much more enriching—students don't have to wonder where to begin, what to write about, or how to compare the differences between areas or countries. The sender and the recipient already have a mutual friend, Flat Stanley. Writing and learning science becomes easier, flows naturally, and tends to be more creative. This is what teachers call an "authentic" science project, in that students are inspired to write about their own passion and excitement for the science project and are given the freedom to write about many sciences through the Flat Stanley character.

Science Activity 6-3 Cartesian Diver vs. Korean HaeNyo (Women Diver)

Materials: A plastic soda bottle with a cap, A pen lid and clay (for weight) or an eye dropper, Water

Activity Procedure (Spangler, 2013a):

Students understand that they are making an air tank of a Cartesian Diver that makes the diver up and down under water.

1. Fill the plastic soda bottle to the very top with water.
2. Fill the glass eyedropper 1/4 full with water. You may need to experiment with the amount of water in the pipette to make it work. Or attach the clay at the end of the pen lid to make the pend lid stand in the water.
3. Place the eyedropper (or pen lid) in the soda bottle. The eyedropper should float, and the water in the bottle should be overflowing. Seal the bottle with the cap (see Figure 6-2).
4. Ask students, "How high will the water level in the eyedropper (or the water level inside the pen lid) be when I squeeze the sides of the bottle?
5. After listening to the students' expectations, then squeeze the sides of the bottle and notice how the eyedropper (called a diver) sinks. Release your squeeze, and it floats back up to the top.
6. Squeeze again and observe the water level in the eyedropper (it goes up).

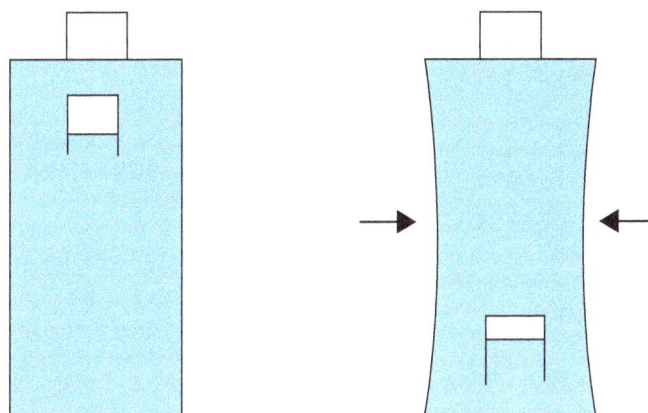

FIGURE 6-2 Cartesian diver experiment

7. Compare with how to make the diver go up and down with Korean women divers (called "HaeNyo"). Provide information about HaeNyo. The students find that they do not have any equipment to make them go up and down inside the water different from Cartesian divers. Then, ask the students, "What makes the HaeNyo go up and down inside the water?"
8. After you listen to their answers, provide the answer, "The HaeNyo use their hands and feet to overcome the water pressure and make them go up to the water surface. Also, to make them go down in the water, they wear an iron belt weighing 15 to 20 pounds.

How it works

"Squeezing the bottle causes the Cartesian diver to sink, because the increased pressure forces water up into the diver, compressing the air at the top of the eyedropper. This increases the mass and density, of the diver causing it to sink. Releasing the squeeze decreases the pressure of the air at the top of the eyedropper, and the water is forced back out of the diver" (Spangler, 2013a). By using the air pressure, the Cartesian divers make them go up and down inside the water overcoming the water pressure. By comparing the Cartesian divers with HaeNyo who use their own bodies to overcome the water pressure, the students understand better how to balance air pressure with water pressure.

Science Activity 6-4 Korean Dumpling, Mandu

Filling Ingredients:

- Kimchi
- Cooked noodles
- Steamed bean sprout
- Tofu
- Ground beef
- Ground pork
- Flower wrappers

Other Materials: Egg, sesame oil, salt, black pepper

Activity Procedure:

1. In a large bowl, mix all ingredients together thoroughly with one egg.
2. Season with 1 tablespoon of sesame oil, 2 teaspoons of salt, and 1 pinch of black pepper. Ensure all the seasoning is mixed in.
3. Place about a tablespoon of #2 filling in the middle of a flower wrapper.
4. Fold the wrapper in half. Seal tightly by pressing the ends together.
5. Boil water in a steamer. Line the steamer with clean cloth; this prevents dumplings from sticking to the steamer. Place dumplings in the steamer, so they are not touching each other. Steam for 15 minutes or until the filling is cooked through.
6. Serve them on a plate with a dipping soy sauce on the side.
7. When the students eat the Mandu, ask them "how many food items does the Mandu have?" The answer will be "The Mandu has grain, vegetables, meat, and bean (except fruit and milk) based on the Food Guide Pyramid of U.S. Department of Agriculture (USDA)."

How it works

When students learn about healthy foods that include all the food items of the Food Guide Pyramid of USDA, the Mandu is a great example for that. Not only the Mandu, but also other foods from other countries can be shared with students during the class.

Science Activity 6-5 Chopstick Game

Materials:
- Chopsticks
- Small sized beans
- Bigger sized beans
- Dishes with two divisions

Activity Procedure:
1. Divide the class into groups.
2. Share the chopsticks among the group members.
3. Have the groups place small-sized beans in one of the two divisions of their dishes.
4. Have the class practice using the chopsticks.
5. Ask the group members move the small beans from one section of the dishes to the other, taking one-minute turns.
6. Do #5 with bigger sized beans.
7. Count the number of beans that each group moves to the other side of the divisions.
8. Give a prize to the winning groups who move the beans more than the other groups.
9. Ask the winner group members how they use the chopsticks when they move the beans, comparing the small and the bigger beans.

How it works

Using chopsticks is applying the "Law of Lever," $d1 \times W1 = d2 \times W2$ (see Figure 6-3), balancing the chopsticks with different weights distributed along its length. When the distance d1 from weight W1 is greater than the distance d2 from the fulcrum to where the weight W2 is applied, then the lever amplifies W1. On the other hand, if distance d1 from the fulcrum to the input force is less than the distance d2 from the fulcrum to the output force, then the lever reduces the input force.

In other words, when the students pick the bigger beans up with the chopsticks, the Law of Lever is applied. The force from the fingers of the students is W1 and the distance between W1 and the imaginary fulcrum of the chopstick is d1. The force from the bigger bean is W2, and the distance between W2 and the imaginary fulcrum of the chopstick is d2. To reduce the force from students' fingers, d1 needs to be longer. Therefore, the groups who understand the Law of Lever will win.

Archimedes said, "Give me a place to stand, and I will move the earth" (Carroll, 2008). He meant that he could lift the Earth if he had a stick long enough to fit the Law of Lever.

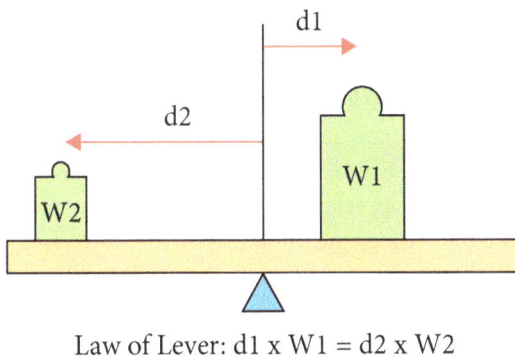

Law of Lever: $d1 \times W1 = d2 \times W2$

FIGURE 6-3 Law of Lever

FIGURE 6-4 Archimedes and Law of Lever

Science Activity 6-6 Tak-ji Game

Materials:

- Two cardboard disks (A4 size)
- Cutting tools (scissors and cutters)

Activity Procedure:

1. Make two Tak-jis to play with (see Figure 6-5):
   a. Fold one cardboard disk in half.
   b. Place it on top of another like cross.
   c. Fold a flap.
   d. Repeat again b and c for the rest of the flaps.
2. Play a Tak-ji game with two players:
   a. Decide who will play first.
   b. The first player throws their Tak-ji disks down on the Tak-jis of the others in an attempt to make the disk flip over.
   c. If a disk is flipped, the player who flipped it gets to keep it.

How it works (Yoon & Martin, 2017)

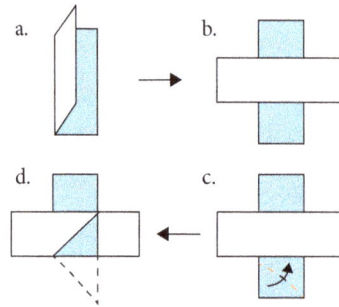

FIGURE 6-5 How to Fold a Tak-ji

Tak-ji is a popular Korean children's' game played using specially made cardboard disks. This game uses Newton's third law, action and reaction. When the first player throws his or her Tak-ji (action), the other player's Tak-ji is flipped (as a reaction). The reaction happens when the force caused by the action is transferred to the other player's Tak-ji appropriately. Through the traditional Korean game, students can understand Newton's third law.

Considerations for Diverse Students

What do you see in the following photo, a young lady or an old woman? You might see a young lady, an old woman, or both. Depending on your view, you will see the same photo differently. Before enter a diverse science classroom, you need to ask yourself as a science teacher to see if you view students differently.

The students in your classroom are diverse, and you need to take care of their diversity. However, the true meaning of taking care of diversity is seeing them as equal. All students in the science classrooms have the knowledge, skills, attitudes, and values necessary to achieve academically, prosper economically, and participate in democracy. Science teachers advocate for every student to receive equitable educational opportunities to pursue their life's goals. The reason why you need to use these various teaching methods to teach them is that the students have rights to learn equally, even though their outside looks different. All of them have passion for learning just like every other. You, as science teachers, need to develop skills to see all of them are the same. This idea develops an equation, "Diversity = Similarity."

Science education must enable all Americans to achieve scientific literacy. Therefore, when teaching various students in science classrooms, science teachers need to consider the following:

Teacher's Learning Style

It is easy to make a mistake, for science teachers to teach students in their own learning styles. For example, when a science teacher is an auditory learner, his or her teaching focuses more on listening and speaking in situations such as lectures and group discussion. Aural teachers use repetition as a teaching technique. To this, students whose preferred learning styles are not auditory but kinesthetic, visual, or

Science Activity 6-7 Kicking a Je-gi

Materials:
- Plastic bags or wrapping tissue paper
- 2 or 3 coins
- String
- Scissors, and
- Scotch tape
- A hacky sack that is same weight with the Je-gi

Activity Procedure:
1. Make a Je-gi:
   a. Pile the coins and tape them together.
   b. Prepare the plastic bag by cutting into a square about 25 centimeters wide.
   c. Place the coin at the middle of the plastic bag.
   d. Hold the coin inside the plastic bag and tie it using a string.
   e. Cut the untied part of the plastic bag into thinner strands (~1.5 centimeters) using scissors.
2. Choose two groups in class: one group plays with the Je-gi, and the other group plays with a hacky sack. The player kicks a Je-gi or a hacky sack up in the air and keeps on kicking to prevent it from falling to the ground. The group with the most consecutive kicks wins. The game rules are:

a. In the Je-gi group, the players stand in a circle and take turns kicking the Je-gi. Players who fail to kick the Je-gi upon receiving it and let it drop to the ground lose. As a penalty, the loser tosses the Je-gi at the winner, so that he or she can kick it as he or she wishes. When the loser catches the Je-gi back with his or her hands, the penalty ends, and he or she can rejoin the game.
b. In the hacky sack group, the same rules are applied.
3. Try #2 with other several groups in class.
4. Compare the results. The class finds that the groups who played with Je-gi win more than the groups with hacky sack.

How it works

Je-gi is a Korean traditional outdoor game object. It requires the use of people's foot and *Jegi*, an object used to play with. Jegi looks like a badminton shuttlecock, which is made of a small coin (quarter size), paper, or cloth. In Korea, children usually play alone or with friends in winter seasons, especially on Lunar New Year.

When playing with a Je-gi that has the same weight with that of hacky sack is compared with a hacky sack, science teachers can talk about air resistance. The surface areas of a Je-gi that experience air resistance are bigger than a hacky sack's. Therefore, the Je-gi stays in the air longer than the hacky sack, and the students can have more time to prepare to kick the Je-gi than the hacky sack (see Figure 6-6).

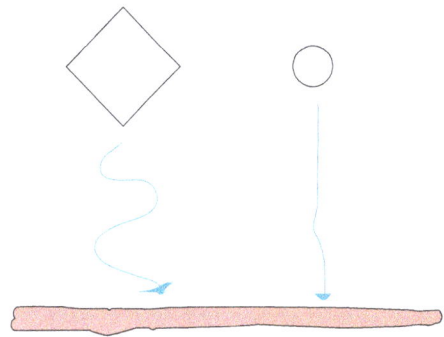

FIGURE 6-6 Comparing the air resistance of a Je-gi with that of a hacky sack

reading/writing may not absorb, process, comprehend, nor retain information in the auditory learning environment. Therefore, it is important for science teachers to understand the differences in their students' learning styles and implement best practice strategies into their diverse learning styles.

Students' Learning Styles

Student learning styles fall into four categories: Visual, Auditory, Reading/Writing Preference, and Kinesthetic (VARK). The VARK model (Fleming & Baume, 2006) "acknowledges that students have different approaches

FIGURE 6-7 Illusion of young lady or old woman

to how they process information, referred to as 'preferred learning modes.'" Students' preferred learning modes have significant influence on their behavior and learning. Therefore, teaching should be matched with appropriate learning strategies and information that is accessed through students' use of their modality preferences shows an increase in their levels of comprehension, motivation, and metacognition. To know students' learning styles, teachers use survey questions and interviews.

5E Learning Cycle

When teachers develop their lessons based on the 5E learning cycle, the lessons are full of inquiry and can cover many students' learning styles. The 5E learning cycles is based on how students learn. It is a research-supported process that can promote more powerful student learning and engagement.

Special Students

Science is for all students. Special students should also be included in science education. Technological aids, like visual and sound impairment applications, are helpful for special students to learn science. For example, Proloquo2go, a symbol-supported communication application, promotes knowledge development and grows communication skills between special students and teachers. Science teachers have ideas of how to integrate technology into their lessons, especially for diverse students, including special students.

Parental Involvement

To take care of diverse students, it is necessary to include parents in the science learning environment. Parents may be the best experts about their children, for example, their characteristics, learning styles, culture, and backgrounds. Therefore, science teachers should prepare by inviting the parents into their classrooms with diverse students, as home instructors, material providers, classroom assistants, or guest speakers. The parents would like to communicate with the science teachers more through hand-written letters or electronic mail. The Parent Teacher Association (PTA) websites provide more ideas of how to get parents involved in classrooms.

Other considerations

Science education in diverse classrooms can be also achieved by including knowledge construction processes, prejudice reduction, and equity pedagogy and by empowering school culture and social structure (Banks, 1993):

- *Knowledge construction process.* Science teachers need to understand the ways in which teachers use activities, methods, and questions to help students understand, investigate, and determine how implicit cultural assumptions, frames of reference, perspectives, and biases within a discipline influence the methods of their knowledge construction.

- *Prejudice reduction*. Teachers create more positive racial and ethnic attitudes to lessen the amount of prejudice among students.
- *Equity pedagogy*. Teachers adapt their instructional methods in ways that
- promote the educational achievement of students from diverse cultural, racial, and gender groups, using different teaching styles.
- *Empowering school and social structure*. Teachers see their schools as complex social systems that must be restructured in order to implement reform related to multiculturalism and diversity.

International Teacher Programs

International teacher programs are another promising opportunity to understand foreign cultures, having teachers interact with another country and learn more about the country's educational culture. During Carlson and Stenmalm-Sjoblom (1992), early childhood teacher candidates from the United States and Sweden visited each other's countries, took coursework, and completed an independent study course. In another program developed by Haley (2012), United States English as a second language (ESL) teacher candidates participated in a 10-week electronic exchange with Chinese English as a foreign language (EFL) teacher education candidates via e-mail and Skype (Yoon & Han, 2016). In other studies, as part of a one-way or two-way teacher exchanges, teacher candidates from the United States taught at schools in a foreign country (Barr, 1995; Fung King Lee, 2011). As a result of their cross-cultural exchanges, teacher candidates have reported they have an increased interest in and understand other cultures through the global teacher programs.

The Global Teacher Internship Program with a Korean university provides a great example that teacher candidates from both America and South Korea have opportunities to interact with public school students' in each other's countries, to teach lessons at schools in each other's countries. The program goals are: a) to acquire knowledge of instructional skills and attitudes to be a successful teacher in the globalized classrooms, b) to understand the educational system and classrooms of each country, c) to interact with people and their cultures, d) to be open-minded to the world, and e) to have flexible ways of thinking.

During their visit abroad, American and South Korean teacher candidates spend the first week of their internship attending education workshops provided by host faculty from participating universities and interacting with several education classes at the partnership university. Next, the teacher candidates are matched with local K–12th grade mentor teachers and are placed in classrooms in elementary, junior high, and high schools for a teaching practicum. The teacher candidates spend 3 weeks in the field. During this time, teacher candidates observe the country's educational system first-hand, and they have opportunities to work with their mentor teachers to develop and to teach several science-related lessons to students in their respective classrooms. After teaching practicum, teachers learn about the host country's culture, the program also includes visits to local museums, to local historical sites, and to a variety of other places of interest.

After participating in the international exchange program, the American and South Korean teacher education students increase their self-efficacy for teaching science. Also, American and South Korean teacher candidates report that the global teacher exchange program increases their global awareness and their ability to be a global teacher.

Visiting world museums, meeting international teachers and students on campus, and inviting local international professionals as guest speakers are also great ways to interact with international people and understand their cultures and education, thus improving global awareness.

Multicultural/Diverse Lessons

To implement multiculturalism/diversity into the lessons, teachers develop skills to modify their existing lessons. Here are the steps:

- Experience scientific concepts through international culture and activities. For example, international students are invited and present their traditional games and show scientific concepts through the games.
- After the introduction to science activities with international students, the teachers are asked to revise their existing lessons by including multicultural/diverse objectives and activities related to the objectives. The teachers also write the rationale to support why their lessons are examples of an "inclusive teaching strategy for a diverse classroom" (see Appendix G).

Ambrosio's four-factor rubric (Ambrosio et al., 2002) is modified and used [to evaluate] the lessons. [The] four-factor rubric is composed of 1) Lesson Plan Objectives, 2) Lesson Plan Mechanics, 3) Lesson Plan Rationale, and 4) Lesson Plan Inclusiveness. Lesson Plan Objectives assess the ability of teachers in designing a meaningful multicultural[/diversity] objectives lesson [indicating proficient] content. Lesson Plan Mechanics evaluates lesson plan activities, assessments, and objectives that match each other. Factor Lesson Plan [Rationales reflect] teachers' respect and affirmations of *individual* differences. Lesson Plan Inclusiveness supports different adaptations to the learning styles of all students. (Yoon, Kim, & Martin, 2016)

Brief explanations of each of the four factors can be found in Appendix G. Each of the factors was scaled from 1 to 4 (1=Incomplete, 2=Unsatisfactory, 3=Developing, and 4= Proficient). Reliabilities for each rubric factor ranged from 0.83 to 0.89. The entire content of the lesson plan and their rationales are considered when scored based on each rubric factor.

The multicultural/diverse lesson is to take care of culturally and linguistically diverse students in science classrooms by adding multicultural/diverse objectives and activities and asking questions for culturally and linguistically diverse students. It is required that science teachers develop multicultural/diverse science lessons for all students.

Summary

All the students in science classrooms have the same fundamental processes for the intellectual growth, but their content and structures vary depending on their backgrounds. Because of the various cultural backgrounds in science classrooms, teachers are required to develop multicultural/diverse lessons by adding multicultural/diverse objectives and activities and find considerations to take care of culturally and linguistically diverse students in advance. The considerations for multicultural education in science classroom are teachers' learning styles, students' learning styles, special students, parental involvement, learning cycles, content integration, knowledge construction processes, prejudice reduction, equity pedagogy, and school culture and social structure empowerment. International teacher exchange programs also provide an opportunity to understand other cultures and visiting world museums, developing multicultural/diverse science lesson, interacting with students from other culture are great ways to experience other cultures.

Assignment

Develop a multicultural/diverse lesson by revising one of your pre-developed lessons. Please follow the direction of Appendix G for this assignment.

References

Ambrosio, A.L., Sequin, C.A., & Hogan, E.L. (2002). "Assessing performance-based outcomes of multicultural/diversity objectives lesson plans: A component within a comprehensive teacher education assessment design." *Multicultural/Diversity Objectives Perspectives*, 3(1), 15–22.

Banks, J. (2013). *An Introduction to Multicultural Education.* Boston: Pearson.

Barr, H. (1995). *International teaching experience and student teacher perspectives of education.* Paper presented at the Annual Meeting of the American Educational Research Association.

Batalova, J. (2006). "Spotlight on Limited English Proficient Students in United States." Migration Policy Institute. Retrieved May 28, 2015 from http://www.migrationpolicy.org/article/spotlight-limited-english-proficient-students-united-states

Carlson, H. L. & Stenmalm-Sjoblom, L. (1992). *Improving teacher education through international cooperation and partnership.* Paper presented at the Meeting of the International Council on Education for Teaching.

Carroll, B. (2008). "Archimedes and the Law of the Lever." Retrieved May 28, 2015 from http://physics.weber.edu/carroll/archimedes/lever.htm

Fleming, N., & Baume, D. (2006). Learning styles again: barking up the right tree!, *Educational Developments, 7*(4), 4–7.

Fung King Lee, J. (2011). "International field experience: What do student teachers learn?" *Australian Journal of Teacher Education, 36*(10), 1–22. Retrieved May 28, 2015 from http://ro.ecu.edu.au/ajte/vol36/iss10/1.

Haley, M. H. (2012). "An online cultural exchange in pre-service language teacher education: A dialogic approach to understanding." *US-China Education Review B, 5,* 528–533.

Hubert, D. (2011). "Flat Stanley: How it works." Retrieved May 28, 2015 from https://www.flatstanley.com/about

Lavatelli, C. (1973). *Piaget's Theory Applied to an Early Childhood Curriculum.* Boston: American Science and Engineering, Inc.

Moore, D. (2014) Cognitive Development. Retrieved from: https://mooredeiante.weebly.com/

Schütz & Kanomata. (2012). "Jean Piaget: Intellectual Development." Schütz & Kanomata company. Retrieved May 28, 2015 from http://www.sk.com.br/sk-piage.html

Spangler, S. (2013a). "Eye Dropper: Cartesian Diver." Retrieved May 28, 2015 from http://www.stevespanglerscience.com/lab/experiments/eye-dropper-cartesian-diver

Spangler, S. (2013b). "Heavy Newspaper." Retrieved May 28, 2015 from http://www.stevespanglerscience.com/lab/experiments/heavy-newspaper-air-pressure-science-experiment

Yoon, J., & Han, I. (2016). "Virtual Activities to Promote Multiculturalism and Sustainability of International Partnerships." IGI Global.

Yoon, J., Kim, K. J., Martin, L. A. (2016). "Culturally inclusive science teaching (CIST) model for teachers of culturally and linguistically diverse students." *Journal for Medical Education, 10*(3). 322-338.

Yoon, J., & Martin, L. A. (2017). "Infusing Culturally Responsive Science Curriculum into Early Childhood Teacher Preparation." *Research in Science Education.*

Yoosma. (2012). "Hand Attacks–Taekwondo Paramus NJ." Retrieved May 28, 2015 from http://www.yoosma.com/content/hand-attacks-taekwondo-paramus-nj

Tip of the Day 6: Sand Animation for Multicultural Science

Hi future science teachers,

Do you remember any game that you played when you were five or six years old? Use your various traditional games to teach science. For example, sand animation can bring an opportunity for students to improve observing, classifying, inferring, predicting, and communicating skills by asking the students to develop stories or make a zoo, a garden, or a local community with sand. You can enjoy the artful sand animation at https://www.youtube.com/watch?v=NkeKAa1u3Lc&feature=youtu.be.

Jiyoon

7

Assessment

How Can You Assess Students in the Science Classroom?

I t was about 8:00 p.m. after dinner. I stopped by a coffee shop before a grocery shopping. When I tried to order a cup of coffee, one of my students was there waiting for my order! I did not know that she worked there. After the order, she pointed and asked a trivia question on the blackboard and waited for my answer. The question was what I had talked about in my science class on that day, "Name one mammal that lays eggs" (see Figure 7-1). Of course, I could answer it ("Echidna!") and received a 10-cent discounted price for the cup of coffee! I really appreciated the opportunity that my student made for me to review what I taught. I will never forget the Echidna!

Through assessment, students can review what they learned and teachers can provide feedback to students on their learning, thus improving students' learning and evaluating effectiveness of teachers' teaching. This chapter illustrates the types of assessment tasks that can be used in the classroom to meet the goals of *A Framework for K–12 Science Education: Practices, Crosscutting Concepts, and Core Ideas* (hereafter referred to as "framework") and the Next Generation Science Standards.

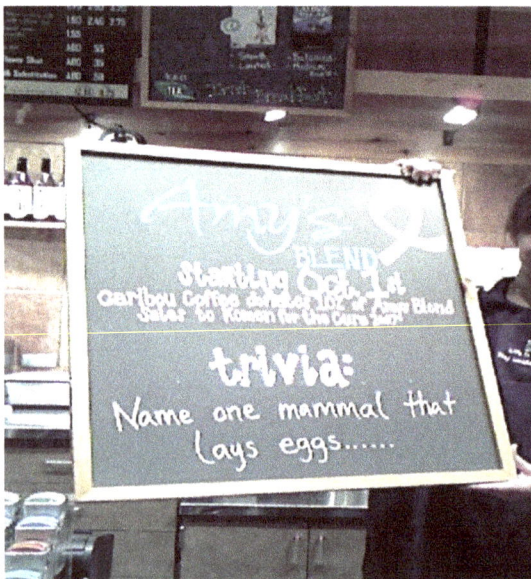

FIGURE 7-1 Trivia question

Science Domains to be Assessed

There are four domains to be assessed in the science classroom:

a. *Inquiry.* Inquiry is one of areas to be assessed in science classrooms. To measure how much inquiry students improve through science learning, science teachers need to develop indicators that provide evidence that a certain condition exists or certain results have or have not been achieved (Brizius & Campbell, 1991). Indicators enable science teachers to assess progress toward the achievement of intended outputs, outcomes, goals, and objectives. Some indicators of student proficiency in inquiry are provided as follows (Olson & Loucks-Horsely, 2000):

- The student engages in scientifically oriented questions.
- The student gives priority to evidence in responding to questions.

- The student formulates explanations from evidence.
- The student connects explanations to scientific knowledge.
- The student communicates and justifies explanations to others.

Each indicator can be rated on a 1 to 4 base (1=not seen; 2 = performed satisfactorily; 3=performed well; and 4=performed in an outstanding and advanced manner), with the average equaling the inquiry grade. The result of the process and inquiry indicator checklists should be coupled with other assessment methods resulting in multiple methods that enable teachers to accurately assess children's progress and achievement in science. Also, it is recommended to develop, together in teams, lists of indicators that tailor to meet the needs of children in classes (Martin, 2012).

b. *Process Skills.* To measure the process skills of students, science teachers can design stations where students can perform their process skills (observing, measuring, predicting, inferring, classifying, communicating). Each station has various ways of measuring process skills and providing process questions, illustrations, physical model, and analogies. A scoring system for the process skill assessment is rated on a 1-to-4 base (1=not seen; 2 = performed satisfactorily; 3=performed well; and 4=performed in an outstanding and advanced manner). Figure 7-2 is a scoring system for process skills assessment. The percentage divided by the maximum total score is a way to indicate students' process skills.

c. *Content.* Science content is frequently the major area to be assessed in schools. The science content can be assessed through lab reports, observation, and written tests. The lab report after an experiments and observation during classes can be scored by a teacher's check-list. The science content is also measured through written tests.

To develop the written test, it is easy for science teachers to follow the matrix of test development provided in Figure 7-3. The written test has two types, selection and supply type:

Scoring System for Practical Process Skill Assessment		
Score	Criterion	
1	Not seen	
2	Performed satisfactorily	
3	Performed well	
4	Performed in an outstanding and advanced manner	
Station	Process Skill	Score
1	Observing	-------------
2	Classifying	-------------
3	Communicating	-------------
4	Measuring	-------------
5	Predicting	-------------
6	Inferring	-------------
7	Predicting	-------------
8	Observing	-------------
9	Classifying	-------------
10	Measuring	-------------
11	Inferring	-------------
	Total	-------------
Divided by maximum total of 44 = percentage -------------		

FIGURE 7-2 Scoring System for Practical Process Skills Assessment

(i) Selection Type
    i. True and False Type
    ii. Multiple Choice
    iii. Matching Items
(ii) Supply Type
    i. Essay Type
    ii. Short Answer Form
    iii. Close or Complete Form

Science teachers first choose the goals of the content, like knowledge, comprehension, or application development. Once they decide the goals, teachers make the difficulties of the question items and choose the types of test items, like selection or supply type.

d. *Attitudes Toward Science*. The attitude toward science is a feeling or way of thinking that affects a student's behavior toward science. It shows how much the student likes science. By comparing students' attitudes toward science before and after they learn science, science teachers can measure students' performance in science.

To measure the attitudes toward science, indicators and surveys are frequently used in science classrooms. Figure 7-4 shows some indicators of attitudes toward science. Each of the positive indicators can be scored by using a 1-to-4 base (1=not seen; 2 = performed satisfactorily; 3=performed well; and 4=performed in an outstanding and advanced manner). Each of the negative indicators is also scored with 1-to-4 scale but in reverse (1 = performed in an outstanding and advanced manner; 2 = performed well; 3 = performed satisfactorily; 4= not seen).

The survey questions to measure the attitudes toward science are developed by using a semantic differential technique, Likert Scale, and observation scale. For example, for the semantic differential technique, an example question is:

Science classes are:

Fun - - - - - - - - - - - Boring

The survey questions can be developed on a Likert scale, for example:

Science is the world's worst enemy.
Strongly Disagree(1), Disagree(2), Neutral(3), Agree(4), Strongly Agree(5).

An example of the observation scaled question is:
Perform the experiment in creative way.

Item type	#	Answer	Credit	Chapter	Contents	Goal			Difficulty		
						Knowledge	Comprehension	Application	Hard	So so	Easy
Selection	1	A	1	Plants	Environment conductive to plant growth	X				X	
	2	C	1	Plants	Essentials for a plant to live		X				X
	3	D	1	Animals	Characteristics of mammals		X			X	
Supply	1	Content	3	Animals	Difference between mammals and reptiles			X		X	
	2	Content	4	Plants	Difference in climate and how it affects plants			X	X		
Total	5		10			1	2	2	1	3	1

FIGURE 7-3 Matrix for Test Development

Authentic Assessment

According to Jon Mueller, "authentic assessment is a form of assessment in which students are asked to perform real world tasks that demonstrate the meaningful application of essential knowledge and skills" (Mueller, 2014, p. 22).

Authentic vs. Traditional Assessment

Authentic assessment is contrary to traditional assessment. When put in an assessment definition continuum, the traditional assessment falls more towards the left end of the continuum and the authentic assessment falls more towards the right end (Mueller, 2014).

Authentic Assessment Tools

A variety of authentic assessment tools are intended to increase students' engagement and make learning more relevant. These include: (1) interview; (2) role play and drama; (3) concept maps; (4) student portfolios; (5) reflective journals; (6) utilizing multiple information sources; and (7) creative group work in which team

Scale	Items
Perception of the science teacher	• Science teachers make science interesting to me. • Science teachers present materials in a way that I understand • Science Teachers are willing to give me individual help.
Anxiety toward science	• When I hear the word "science," I have a feeling of dislike." • It makes me nervous to even think about doing science. • I have a good feeling toward science
Value of science in society	• Science is helpful in understanding today's world. • Science is of great importance to a country's development. • There is little need for science in most of today's jobs.
Self-concept in science	• No matter how hard I try, I cannot understand science. • I do not do very well in science. Science is easy for me.
Enjoyment of science	• Science is something that I enjoy very much. • I enjoy talking to other people about science. • Science is one of my favorite subjects.
Motivation in science	• I would like to do some extra or un- assigned readings in science. • The only reason I am taking science is because I have to. • I have a real desire to learn science.
Attitude towards science in school	• When I hear the word "science," I have a feeling of dislike. • During science class, I usually am interested. • I would like to learn more about science. Science makes me uncomfortable, restless, irritable, and impatient. • Science is fascinating and fun. • I feel a definite positive reaction to science. • Science is boring.

FIGURE 7-4 Some indicators of attitudes toward science (Desy, Peterson, Brockman, 2011) *Note:* The first six scales were derived from Gogolin and Swartz (1992); the 'Attitude toward science in school' scale was derived from Germann (1988). Reverse coding was used to quantify responses on negatively worded items.

members design and build models by using drawing, drama, and construction (Aitken & Pungur, n.d., 3).

• Interview is one of the best ways to find out how much children have learned and how well they understand what they have learned. Open-ended and partially structured interviews with individual children accomplish this goal and are truly

Science Activity 7-1 State, National, and International Tests

Materials: Results from state, national, and international tests

Activity Procedure:

1. Ask the class to make groups (5 or 6 members in a group)

2. Find state, national, and international tests online and discuss what to test more to improve student performance in science

3. Develop at least ten of your own test questions to measure students' performance in science.

4. Share them with other groups.

authentic ways of obtaining information about children's achievement and their thinking.

- Writing a journal or short science story is another way to assess student performance. Journal writing is to enhance reflective learning and to deepen critical thinking about earning (e.g., Andrusyszyn & Davie, 1997; Boud, 2001).
- When teachers use student portfolios as authentic assessments, selection, reflection, and

mechanics are the rubrics to assess the student portfolio. A students' portfolio is graded depending on how well the student selected his or her artifacts representing his or her understanding, how much the he or she reflected on his or her learning, and how well he or she developed the artifacts.

- A concept map is a diagram showing the relationships among concepts. The rubric for the

Traditional --------------------------------- Authentic
Selecting a Response -------------------------- Performing a Task
Contrived --- Real Life
Recall/Recognition --------------------- Construction/Application
Teacher-Structured ------------------------- Student-Structured
Indirect Evidence ---------------------------- Direct Evidence

FIGURE 7-5 Continuums of attributes of traditional and authentic assessment

Selecting a Response vs. Performing a Task: On traditional assessments, students are typically given several choices (e.g., a, b, c, or d; true or false; which of these match with those) and asked to select the right answer. In contrast, authentic assessments ask students to demonstrate understanding by performing a more complex task, usually representative of more meaningful application.

Contrived vs. Real Life: It is not very often in life outside of school that we are asked to select from four alternatives to indicate our proficiency at something. Tests offer these contrived means of assessment to increase the number of times you can be asked to demonstrate proficiency in a short period of time. More commonly in life, as in authentic assessments, we are asked to demonstrate proficiency by doing something.

Recall/Recognition of Knowledge vs. Construction/Application of Knowledge: Well-designed traditional assessments (e.g., tests and

quizzes) can effectively determine whether or not students have acquired a body of knowledge. Thus, as mentioned above, tests can serve as a nice complement to authentic assessments in a teacher's assessment portfolio. Furthermore, we *are* often asked to recall or recognize facts and ideas and propositions in life, so tests are somewhat authentic in that sense. However, the demonstration of recall and recognition on tests is typically much less revealing about what we really know and can do than when we are asked to construct a product or performance out of facts, ideas, and propositions. Authentic assessments often ask students to analyze, synthesize, and apply what they have learned in a substantial manner, and students create new meaning in the process, as well.

Jon Mueller, Excerpt from: "What is Authentic Assessment?," http://jfmueller.faculty.noctrl.edu/toolbox/whatisit.htm. Copyright © by Jon Mueller. Reprinted with permission.

Teacher-Structured vs. Student-Structured: When completing a traditional assessment, what a student can and will demonstrate has been carefully structured by the person(s) who developed the test. A student's attention will understandably be focused on and limited to what is on the test. In contrast, authentic assessments allow more student choice and construction in determining what is presented as evidence of proficiency. Even when students cannot choose their own topics or formats, there are usually multiple acceptable routes towards constructing a product or performance. Obviously, assessments more carefully controlled by the teachers offer advantages and disadvantages. Similarly, more student-structured tasks have strengths and weaknesses that must be considered when choosing and designing an assessment.

Indirect Evidence to Direct Evidence: Even if a multiple-choice question asks a student to analyze or apply facts to a new situation, rather than just recall the facts, and the student selects the correct answer, what do you now know about that student? Did that student get lucky and pick the right answer? What thinking led the student to pick that answer? We really do not know. At best, we can make some inferences about what that student might know and might be able to do with that knowledge. The evidence is very indirect, particularly for claims of meaningful application in complex, real-world situations. Authentic assessments, on the other hand, offer more direct evidence of application and construction of knowledge. As in the golf example above, putting a student on the golf course to play provides much more direct evidence of proficiency than giving the student a written test. Can a student effectively critique the arguments someone else has presented (an important skill often required in the real world)? Asking a student to write a critique should provide more direct evidence of that skill than asking the student a series of multiple-choice, analytical questions about a passage, although both assessments may be useful.

Source: Mueller, 2014

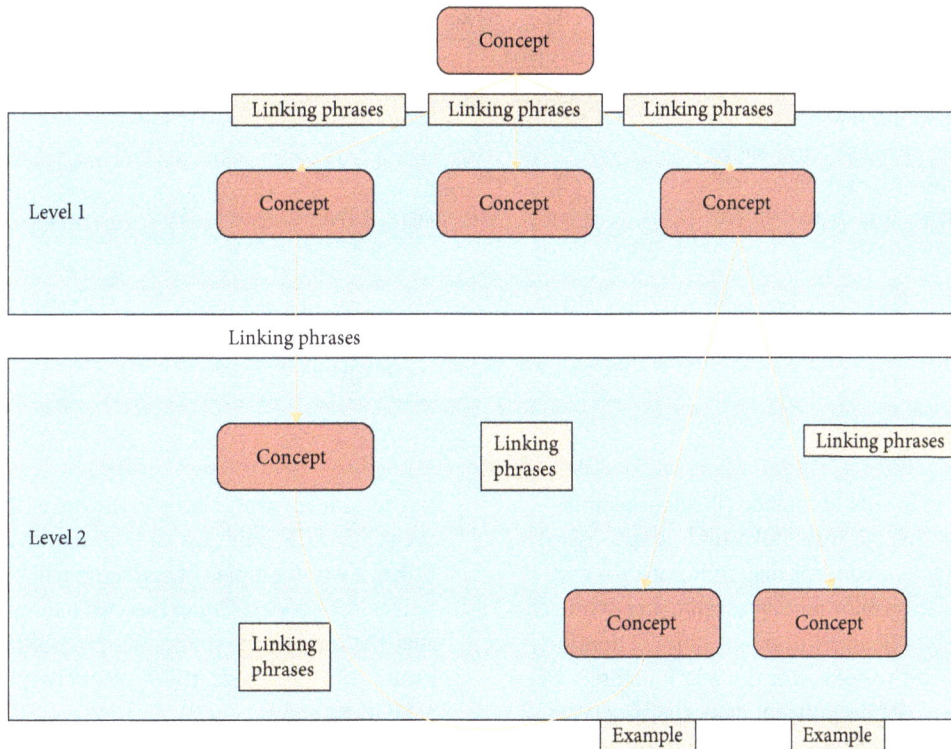

FIGURE 7-6 A chart to show how to grade a concept map

Science Activity 7-2 Assessment with a Concept Map

Materials: Two concept maps

Activity Procedure:

1. Provide scores for Jenny's concept maps when she was in grade 2 (Figure 7-7) and 12 (Figure 7-8). Rubric for concept map is (1) relationships (1 point); (2) hierarchy (5 points); (3) cross links (10 points); and (4) examples (1 point).

2. Based on scores, explain how much Jenny understands the concepts and how much she has improved her learning by comparing the two concept maps.

"Authentic assessment provides a measure by which student[s'] academic growth can be gauged over time while capturing the true depth of learning and understanding. It moves beyond the practices of traditional tools and tasks and allows for a greater expression of students' abilities and achievement" (Aitken & Pungur, n.d., 3).

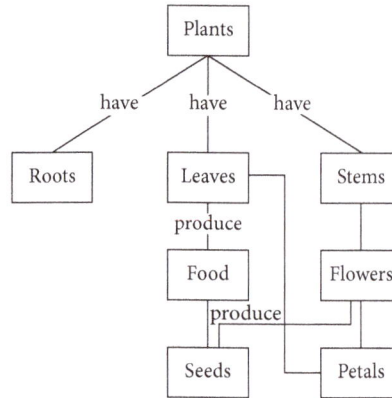

FIGURE 7-7 Jenny's concept map in grade 2

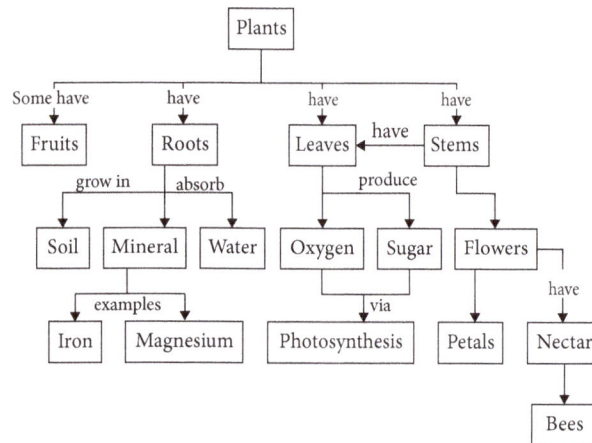

FIGURE 7-8 Jenny's concept map in grade 12

concept map should include (1) relationships; (2) hierarchy; (3) cross links; and (4) examples. For example, a concept map in Figure 7-6 has seven relationships among seven concepts. Each of the relationships will have 1 point. Therefore, the total of the points for the relationship is 7 (= 1 point × 7). In the concept map, there are two levels so that the total of the hierarchy points is 10 (= 5 points × 2 levels). The cross-links that link between phrase links will have a high score.

In the concept map, there is one cross-link and the total of the points is 10 (= 10 points x1 cross-links). Also, examples of concepts will be given points. The concept map has two examples so that the total of the points is 2 (= 1 point x 2 examples). Therefore, this concept map has a total of 29 points.

A concept map is used for (1) diagnostic, formative, and summative assessment; (2) curriculum and lesson development; and (3) a science teaching and learning method. Before, during, and after a lesson, teachers ask students to develop a concept map to assess their learning. Also, a concept map enables teachers to have ideas of how to create and organize lessons and curriculum. By looking at students' concept maps, science teachers understand how much they need to improve their teaching and which areas the students need more support.

Rubrics

Rubrics can be used to assess almost any type of student work, for example,

> essays, final projects, oral presentations, or theatrical performances. They can be used at the time an assignment is given to communicate expectations to students when student work is evaluated for fair and efficient grading and to even assess a program by determining the extent to which students are achieving learning outcomes. A rubric is simply a scoring tool that identifies the various criteria relevant to an assignment or learning outcome, and then explicitly states the possible levels of achievement along a continuum (poor to excellent or novice to expert). ("Rubrics," n.d.)

Rubrics make grading easier and faster in several ways (Stevens & Levi, 2004a):

- Establishing performance anchors
- Providing detailed, formative feedback (three- to five-level rubrics)
- Supporting individualized, flexible, formative feedback (scoring guide rubrics)
- Conveying summative feedback (grade)

In constructing a rubric, four basic stages are involved (Stevens & Levi, 2004b):

- *Stage 1: Reflecting.* Teachers take the time to reflect on what they want from the students, why

they created this assignment, what happened the last time they gave it, and what their expectations are

- *Stage 2: Listing.* Teachers focus on the particular details of the assignment and what specific learning objectives they hope to see in the completed assignment.
- *Stage 3: Grouping and Labeling.* Teachers organize the results of their reflections in Stage 1 and 2, grouping similar expectations together in what will probably become the rubric dimensions.
- *Stage 4: Application.* Teachers apply the dimensions and descriptions from Stage 3 to the final form of the rubric.

Science Report Card

Inquiry-based science teachers need to consider all the areas of science education, including inquiry, process skills, content, and attitudes, to be assessed when they develop science report cards for their students. Some of them might think there is no provision for content achievement. However, in today's climate of standardized tests and high-stakes accountability, teachers may find it necessary and desirable to include a percentage for content achievement.

Science report cards should be calculated in accordance with a weighted system that has been pre-established and, especially in the case of upper elementary grades, shared with the children. Either the point system or the percentage system can be employed. A typical plan for establishing marks (point or percentage system) might look like that portrayed in Table 7-1.

It is difficult to assign marks in process-oriented inquiry science, because it is difficult to establish measurable criteria that can be administered uniformly with all children. However, suggestions provided in this chapter for quantifying assessment information and preparing marks help teachers grade their students both fairly and consistently. Above all, teachers' report cards should be positive, reflect the progress of individual children, and communicate individual success to both children and their parents. As a conclusion,

Science Activity 7-3 Prepare Teacher License Test

Assessing student performance is another way to assess teachers' teaching. After reviewing the report card, teachers need to support what the students have missed in their science learning and develop instructional strategies accordingly. The test for the teacher license test is meant to demonstrate teacher quality in states. The teacher license test may be a reliable assessment tool to measure the knowledge and skills of teacher-candidates. Appendix J provides materials that are helpful for teacher-candidates to prepare the teacher license tests in states.

TABLE 7-1 An Example of a Science Report Card.

Areas (%)	Inquiry (50%)		Process (30%)	Contents (10%)	Attitude (10%)	Total (100%)	A–F
Methods	Worksheet	Journal	Science fair	Q/A	Survey		
Score							
Susan Lee							
Jean Smith							
Tom Johnson							
William Dish							

the report card represents the teacher's philosophy of teaching for maximum individual growth (Martin, 2014).

Summary

To adequately cover a framework for K–12 Science Education (Scientific/Engineering Practices, Crosscutting Concepts, and Core Ideas), the Next Generation of Science Standards (NGSS) positions that assessment tasks use a variety of response formats in the areas of inquiry, process skills, content, and attitudes. Authentic assessment asks students to perform real world tasks that demonstrate the meaningful application of essential knowledge and skills, including interviews, portfolios, role-play and drama, concept maps, reflective journals, utilizing multiple information sources, and creative group work using drawing, drama, and construction. Rubrics "can be used at the time an assignment is given to communicate expectations to students, when student work is evaluated for fair and efficient grading, and to even assess a program by determining the extent to which students are achieving learning outcomes" ("Rubrics," n.d.). Quantifying assessment information and preparing points or percentage systems helps teachers grade their students both fairly and consistently. Teachers need to understand that their report cards should be positive, reflect the progress of individual children, and communicate individual success to both children and their parents. Teachers need to understand that the report card represents the teacher's philosophy of teaching for maximum individual growth.

Assignments

Develop your own science report card of a student, including science assessment methods (at least two different methods for each area of inquiry, process skill, attitudes, and content) to measure inquiry, process, attitudes, and the content of science. In all the science domains to be assessed, you need to consider a framework for K–12 Science Education (scientific/ engineering practices, crosscutting concepts, and core ideas) in your report card.

References

Aitken, N. & Pungur, L. (n.d.). *Authentic Assessment.* 1-8. Retrieved from: http://www.ntu.edu.vn/Portals/96/Tu%20lieu%20 tham%20khao/Phuong%20phap%20danh%20gia/authentic%20 assessment%202.pdf

Andrusyszyn, M. A. & Davie, L. (1997). "Facilitating Reflection through Interactive Journal Writing in an Online Graduate Course: a Qualitative Study." *Journal of Distance Education,*12, 103–126.

Boud, D. (2001). "Using Journal Writing to Enhance Reflective Practice." In English, L. M. and Gillen, M.A. (Eds.) *Promoting Journal Writing in Adult Education* (New Directions for Adult and Continuing Education, No. 90, pp. 9–18). San Francisco: Jossey-Bass.

Brizius, J. A., & Campbell, M. D. (1991). *Getting results: A guide for government accountability.* Washington, DC: Council of Governors Policy Advisors.

DePaul University. (2014). "Types of Rubrics." Retrieved June 6, 2015 from http://teachingcommons.depaul.edu/Feedback_ Grading/rubrics/types-of- rubrics.html

Desy, E., Peterson, S., & Brockman, V. (2011). "Gender Differences in Science-Related Attitudes and Interests Among Middle School and High School Students." *Science Educators, 20*(2), 23–30.

King, P.M. & Baxter Magolda, M.B. (2005). "A developmental model of intercultural maturity," *Journal of College Student Development,* 46(2), 571–592.

Martin, D. (2012). *Elementary Science Methods: Constructive Approach.* Belmont, CA: Wadsworth.

Mueller, J. (2014). "What is Authentic Assessment?" Retrieved June 6, 2015 from http://jfmueller.faculty.noctrl.edu/toolbox/ whatisit.htm

National Research Council. (2013). *Report Brief: Developing Assessments for the Next Generation Science Standards.* Board on Tests and Assessment, Board on Science Education. Retrieved June 6, 2015 from http://sites.nationalacademies.org/ cs/groups/ dbassesite/documents/webpage/dbasse_086205.pdf

Olson, S. & Loucks-Horsley, S. (Eds.). (2000). *Inquiry and the National Science Education Standards: A guide for teaching and learning.* Washington, DC: National Academy Press. Retrieved June 16, 2015 from http://www.nap.edu/books/ 0309064767/ html/ or http://books.nap.edu/html/inquiry_addendum/

Rubrics. (n.d.). DePaul University Office for Teaching, Learning, and Assessment. Retrieved from: https://resources.depaul.edu/ teaching-commons/teaching-guides/feedback-grading/rubrics/ Pages/default.aspx

Stevens, D. & Levi, A. (2004a). Grading with Rubrics. *Introduction to Rubrics: An Assessment Tool to Save Grading Time, Convey Effective Feedback and Promote Student Learning.* Sterling, VA: Stylus Publishing. 73–76.

Stevens, D. & Levi, A. (2004b). "How to Construct a Rubric." *Introduction to Rubrics: An Assessment Tool to Save Grading Time, Convey Effective Feedback and Promote Student Learning.* Stylus Publishing. 29–46.

Tip of the Day 7: Show Your Teaching Philosophy

Hi future science teachers,

Assessing students is to show your teaching philosophy. Many of you knew about Albert Einstein who *failed* math when he was young. Thomas Edison's teacher declared that he was mentally retarded and gave up teaching him at the school. I do not want you to make such an assessment in your science classroom. Think about what is the most important thing in your teaching. Ask yourself, "Why do you teach?" Based on the answer, assess your students.

My answer to the question is that "All of my students have their own talents in science. I will help them to develop their own talents in science." I create various assignments to measure students' learning in science. Even though the students cannot make some of the assignments because of their own situations (like a busy day, a bad physical condition, and so on) and different learning styles, they can make up their points from other assignments. I do not lose another Edison in my science classroom!

Jiyoon

8 Management and Safety

How Can You Manage Your Science Classroom with Safety?

Whenever I tried to talk about lab safety, I came in the class wearing my snowboard goggles and graduation gown on me and asked the class if I looked safe in a science lab. Then, most of the students pointed out my gown because the loose sleeves can easily catch fire over an oven. Interestingly, they never talk about the snowboard goggles that I wore, because they already understood that the snowboard goggles meant safety. So, I decided to use my photo with the snowboard goggles as a safety poster to remind the students of safety in a lab. Since I posted the poster on the wall, it has been easy for me to guide my students in safety. Whenever the students had an experiment in science classrooms, all I had to do was point to the safety poster and ask them to wear googles and be safe! Figure 8-1 shows the poster with two different kinds of goggles: Snowboard goggles on my head and the lab goggles in the sterilizing storage cabinet.

This chapter presents strategies of how to manage a science classroom keeping the classroom safe. Maslow's hierarchy of needs, family involvement, classroom environment, and safety guidelines will be presented through this chapter.

FIGURE 8-1 A safety poster with two different kinds of goggles

Maslow's Hierarchy of Needs

"Keira Wilmot was a typical 16-year-old honor student at a high school in Florida. She's never been in trouble. She has a reputation for being nice to everyone, getting straight A's, and loving science" (Welsh, 2013). But she was arrested as an adult for allegedly possessing and discharging a weapon on school grounds and discharging a destructive device for doing a science experiment, pouring toilet cleaner and aluminum foil inside a plastic bottle and making an explosion (Welsh, 2013). Many science educators have argued about the decision of this accident but the most important lesson from the accident was safety. It is strongly recommended that science teachers need to keep a safe science environment all times, more than for creative experiments.

To keep science classrooms safe, teachers may tend to teach in ways that ensure good order, for example, recitation, teacher-dominated questions and answers, worksheets, choral reading, and so on. Teaching discovery and inquiry modes run the risk of disorder, and only the most courageous teachers are willing to take that risk. However, the primary rule in a constructivist classroom is that every child has the right to learn and the teacher has the right to facilitate this learning. Teachers need to have skills to balance regulations and prescribed punishment for given infractions and positive rewards for desired behaviors in maintaining classroom order.

Psychologist Abraham Maslow proposed the idea that everyone seeks to satisfy five basic kinds of needs: physiological, safety, belonging, esteem, and self- actualization:

- Tier 1: Physiological: Hunger, Thirst, and Sex
- Tier 2: Safety: Security and Stability, Dependency, Protection and Freedom from Fear
- Tier 3: Social: Affection, Belongingness
- Tier 4: Esteem: Strength, Reputation, Confidence, Importance
- Tier 5: Actualization: Self-Fulfillment, Cognitive Needs

The most elemental of an individual's needs are physiological: food, drink, shelter, and sexual satisfaction. The second step up the hierarchy is the need for safety, shelter, and protection from physical and emotional harm. The third level of needs is social ones—for love and a sense of belonging from parents, siblings, or extended families. The next level of need is for esteem. Needs here are both internal (for example, self-respect and autonomy) and external (for example, status and attention). Finally, an individual works to gain self-actualization, which is a "knowing" about life and its meaning for the individual and a sense that she or he fits into the paradigm (Sharlyn, 2005).

Students do not all come to school with the same resources having been met in terms of basic necessities. When each tier of needs has been met, there is motivation for meeting and acquiring the next. The classroom activities that science teachers can do to satisfy students' needs include some of the following:

Tier 1: Physiological Needs

Water bottles, water breaks, food, and instrumental music provide necessities to the students in classrooms. Also, physical surroundings, including room arrangement, color, temperature, and plants, are related to ensure the needs of students in classrooms are met (Desautels, 2014).

Tier Two: Stability, Safety and Security, Freedom from Fear

A worry drop box frees up working memories and relieves anxiety by writing out students' concerns and worries. There are many other ways to develop classroom safety, like making class safety guidelines together, creating a class blog, and inviting outside speakers that promote safety and security, like police officers, counselors, former students who have risen above difficult situations, etc. (Desautels, 2014).

Tier Three: Belonging and Love

Classroom service projects and partnered work can fulfill this need, like creating special and celebratory days all year long and working together by assigning the roles within the class. Also, students share movie clips, personal narratives, or a story at the beginning of the day and develop a classroom theme, flag, song, flower, and animal totem as a group (Desautels, 2014).

Tier Four: Achievement, Recognition and Respect of Mastery, Self-Esteem

For students to feel capable and successful, science teachers must create an environment where students get to demonstrate personal expertise, for example, a career day by bringing college students and community professionals to share the possibilities of academic and professional success (Desautels, 2014).

Tier Five: Self-Actualization and Self-Fulfillment Needs

Individual students may perceive or focus on this need very specifically. For example, an individual student may have the strong desire to become an ideal parent. In another, the desire may be expressed athletically. For others, it may be expressed in paintings, pictures, or inventions. "In this tier, students become self-assessors and self- reflectors. They are able to see and understand how their actions, thoughts, and feelings affect all lives" (Desautels, 2014).

Family Involvement

According to the National Science Teachers Association (2009a),

the involvement of parents and other caregivers in their children's learning is crucial to their children's interest in and ability to learn science. Research shows that when parents play an active role, their children achieve greater success as learners, regardless of socioeconomic status, ethnic/racial background, or the parents' own level of education (PTA 1999; Henderson and Mapp 2002; Pate and Andrews 2006). Furthermore, the more intensely parents are involved, the more confident and engaged their children are as learners and the more beneficial the effects on their achievement (Cotton and Wikelund 2001).

Culturally and linguistically diverse children need parental involvement in science rooms. According to Key (2010), many children of color have been found to have an external locus of control. *Locus of control* is the concept or belief about the source of one's fate and destiny that is a powerful predictor of academic achievement (Brookover, et. al., 1979; Das & Pattanaik, 2013; Madu, 2018). A student with an internal locus of control believes personal successes or failures are due largely to his/her own abilities and efforts. A student with an external locus of control believes their successes or failures are due largely to external factors, luck, other people's actions, or difficult situations. Minority children who have experiences at school that are different from what they have had at home cannot connect easily their learnings at schools to constructional knowledge and, thus, cannot achieve their performance successfully, which makes them lose confidence in themselves.

Parents are the key individuals who can facilitate the learning environment for their children by converting to an internal locus of control. Parents can provide children with positive feedback and encourage them to suggest various ways of investigating a given phenomenon in their daily home lives (Curtis-Fields, 2010). The parents can employ their own culture to develop a friendly environment and have the children explore more. Many children are comfortable interacting with people, behaviors, and ideas that they are familiar with but react with fear and apprehension when faced with the unfamiliar (Alsubaie, 2015). The respectful and comfortable learning environment enables children to investigate and set their own goals and evaluate their own progress with freedom. The parents can develop the learning environment in which the children trust their judgment, become more independent, and establish an internal locus of control. According to Parent and Family Involvement Survey (PIF) of the National Household Education Surveys 2007 (Hagedorn et al., 2008), there has been a significantly positive influence of parental empowerment on their culturally diverse children when their culturally sensitive and culturally unique empowerments were involved. Many studies on ethnic minority students have documented the positive impact of culture on students' achievement (Kim, 2012; Kim, 2002; Kim & Rohner, 2002).

The National Standards for Parent/Family Involvement Programs and their quality indicators are research based and grounded in both sound philosophy and practical experience. The purpose for the standards is threefold (National PTA, 2014):

- To promote meaningful parent and family participation
- To raise awareness regarding the components of effective programs
- To provide guidelines for schools that wish to improve their programs

TABLE 8-1 **National Standards for Parent/ Family Involvement Programs.**

National Standards for Parent/Family Involvement Programs
Standard I: Communicating Communication between home and school is regular, two ways, and meaningful.
Standard II: Parenting Parenting skills are promoted and supported.
Standard III: Student Learning Parents play an integral role in assisting student learning.
Standard IV: Volunteering Parents are welcome in the school, and their support and assistance are sought.
Standard V: School Decision Making and Advocacy Parents are full partners in the decisions that affect children and families.
Standard VI: Collaborating with Community Community resources are used to strengthen schools, families, and student learning

The National Standards for Parent/Family Involvement Programs were developed by the National PTA in cooperation with education and parent involvement professionals through the National Coalition for Parent Involvement in Education (NCPIE). Building upon the six types of parent involvement, these standards, together with their corresponding quality indicators, were created to use in conjunction with other national standards and reform initiatives in support of children's learning and success (National PTA, 2014).

TABLE 8-2 **Common types of parent involvement and relevant indicators.**

Common Types of Parent Involvement	Indicators
Parent participation at school	Parent participation in school activities Parent helps in child's classroom
Parent involvement at home	Parent reads to child Parent cooks with child Parent goes on outings with child Parent and child take trips to other cities Parent take trips to the zoo
Parent involvement in school and learning	Parent helps child with homework Parent makes sure child does homework Parent provides learning experiences for the child Parent communicates with the school regularly Parent talks to child's teacher Parent communicates with the school Parent discusses school progress with child Parent and child discuss school at home Parent pick up child's report cards
Parent participation in community activities	Parent attends local school council meetings Parent votes in school elections Parent is a member of a community organization Parent is a member of PTA or other school group

Parent involvement can take many forms. Common indicators include home support for learning, parenting practices, child–parent interactions, participating in school activities, involvement in school association, involvement in school governance or in community activities, and expectations for children's success or educational attainment (see table 8-2) (Christenson & Reschly, 2010).

Classroom Environment

To manage science classrooms, science teachers need to pay attention to create a classroom environment where students satisfy their needs, composed of the following components:

- *Oxygen-Rich*. "Oxygen is essential for brain function, and enhanced blood flow increases the amount of oxygen transported to the brain. Physical activity [in classroom] is a reliable way to increase blood flow, and hence oxygen, to the brain" (Anderson, Eckburg, & Relucio, 2002). Science teachers make opportunities for the children to go outside and take in fresh air every day and also have more oxygen in their brains through regular physical activities.
- *Animals and Plants*. Live plants and animals in the classroom can be wonderful learning opportunities for students. More than just decorations, these live plants and animals can turn a classroom into a center for observing, questioning, data collecting, and developing a respect for living things.

According to the National Science Teachers Association (2009b):

Before teachers start, they should check out district or school policies or guidelines for live plants and animals. Also, teachers need to consider their science curriculum and standards, thinking about what learning goals are supported by having plants and animals in the classroom. Rather than relegating the animals to the back of the classroom and the plants to the windowsill, creating a learning center that can focus students' attention with questions (especially student-generated ones) and related resources.

Teachers need to be careful in choosing animals. Venomous animals are not recommended to have in science classrooms because of distracting amounts of noise and their requiring controlled environments (as some reptiles do). Also, teachers need to find out if any students have allergies to hair, fur, or feathers. Wild animals, such as chipmunks or songbirds, also do not belong in the classroom (and possessing them may be in violation of state or local game laws). Small rodents, such as guinea pigs, mice, or hamsters are popular classroom residents. Also, hissing cockroaches, snakes (such as ball pythons or corn snakes), and others (such as bearded dragons, iguanas, turtles, or tree frogs) are suggested as classroom animals. It is suggested that you get animals from a reputable pet shop or other provider (including rescue organizations) that can advise the teachers and the students on their housing and care.

Aquariums are also popular in classrooms. Students (and teachers) find them relaxing and interesting to observe. If teachers and students have never set one up before, it is recommended they get a small "starter kit" and some inexpensive tropical fish. It would be a great experience for teachers and their students to learn together.

There are many opportunities for inquiry with plants, especially if students start them from seeds or clippings. Teachers should choose plants that do not have poisonous leaves or berries for science classrooms. It would be better to have a small houseplant for each student in class. The students can have chances to decorate the pots and take them home at the end of the year if they wish.

Science Activity 8-1 Film Canister Rocket Launch Competition for safety

Materials: See the items in the handout for the film canister rocket launch competition (see Table 8-2).

Activity Procedure:

1. Ask the class to make groups (five or six members in a group)
2. The teacher will be a seller of the materials for the film canister rocket competition and ask the groups to purchase the materials to make film canister rockets.
3. Before purchasing the materials, have the groups discuss which materials they need to buy to develop film canister rockets with the best budget.
4. After they develop their own rockets with the materials they purchased, go outside and take the time to launch each rocket.
5. The group that builds a rocket that flies the highest for the lowest expense will be the winner in this competition.
6. Have the groups discuss the last question from the handout.

How it works:

This activity is a good way to talk about not only chemical reaction but also safety. The white film canister works better than the black canister, because the lid of the white canister does not need more power to open than the black one. The gas made by the chemical reaction of the Alka-Seltzer tablet with water inside the film canister is the resource for the power to open the lid. When the lid is open, the rocket is launched into the air. Therefore, the more gas made, the higher the rocket is launched. The lid of the film canister needs to seal the canister well enough to hold the gas until it explodes open! Science teachers need to ask the class to wear goggles for safety when the rockets are launched.

TABLE 8-3 **Handout for the Film Canister Rocket Launch Competition.**

ITEMS	PRICE
Paper	$1/sheet
Scissors	$5/pair
Alka-Seltzer tablet	$5/a quarter of a tablet (one tablet limitation)
Film canister	$5 (black), $10 (white)
Water	$5/one third of the canister
Blue chip: $10, Red chip: $5, White chip: $1	

The group builds a rocket that flies the highest for the least money will be the winner in this competition!
Here is the guidance:

1. Discuss in groups the materials to buy to make a rocket with the best budget. The rocket will be launched by the carbon dioxide from the Alka-Seltzer. List the items that you want to buy and their prices.
2. Buy the materials and build your rocket. Total Price: _____
3. At the launch mark, fill the canister with water and add the Alka-Seltzer tab. Quickly place the lid on the canister; make sure it's on tight and place—lid side down—on a flat surface. Stand back, and watch.

Discuss, in groups, the question, "Which film canister works better? And why?"

Science Centers (Bigelow, 2014)

Some classroom science centers (also called science stations or tables) include activities for students to do on a rotating basis. This is useful when there are not enough materials for an entire class, when teachers want to provide a choice of activities, or for providing alternative or more advanced activities for interested students. These centers include directions, and the activities should be safe enough for students to do independently. Teachers should have procedures in place for how and when students access the center.

Creating centers for each unit can be time-consuming for the teacher, especially during the first year An alternative is to have students contribute to them, giving students ownership of the project. The science centers help students create and sustain the appropriate environment in which to build a community of active, inquisitive learners. A well-displayed, interactive science center appeals to students' sense of curiosity and promotes interest, discussion, and research.

There are no hard and fast rules, and the size or complexity of teachers' "corner" depends on the space the teachers have. Even a small table can be effective. The science centers give students opportunities to contribute and make suggestions.

Culturally and Linguistically Integrated Learning Environment

As science classrooms become culturally and linguistically diverse, teachers need to manage classrooms, including content integration, knowledge construction processes, prejudice reduction, equity pedagogy, and empowering school culture and social structure.

- Content Integration. Science teachers use examples and content from a variety of cultures and groups to illustrate key concepts, principles, generalizations, and theories.
- *Knowledge construction process.* Science teachers need to understand the ways in which teachers use activities, methods, and questions to help the culturally and linguistically diverse students understand, investigate, and determine how implicit cultural assumptions, frames of reference, perspectives, and biases

within a discipline influence their methods of knowledge construction.
- *Prejudice reduction.* Teachers create more positive racial and ethnic attitudes to lessen the amount of prejudice among the culturally and linguistically diverse students.
- *Equity pedagogy.* Teachers adapt their instructional methods in ways that promote the educational achievement of students from diverse cultural, racial, and gender groups, using different teaching styles.
- *Empowering school and social structure.* Teachers see their schools as complex social systems that must be restructured in order to implement reform related to multiculturalism and diversity.

Science is for all the diverse students in the classroom. The students have rights to learn science and the science teachers are the guides to the students. For more strategies to manage culturally and linguistically diverse science classrooms, teachers should see Chapter 6.

Safety in the Science Classroom (American Chemical Society, 2014)

When managing science classrooms, safety is one of the most important considerations science teachers must keep in mind. Schools provide many ways to make the classroom safe.

The American Chemical Society (ACS) suggests the following directions (see adjacent box) for safety in the elementary science classroom (American Chemical Society, 2014):

Before the Experiment

Many potential hazards can be eliminated if the teacher has an organized and disciplined classroom. To do this, the teacher needs to perform the experiment before assigning it to the students. Then, as a result of the prior performance, the teacher will be familiar with the activity, will have the materials ready to distribute to the students, will be ready to supervise the students' activities, will have a plan for collecting materials after the activity, and will be able to instruct the students in what is expected of them.

Eye and Personal Protection

1. Teachers should always wear chemical splash-proof safety goggles when working with chemicals, as should students working or watching in the area. Child-sized goggles are available from science materials suppliers.
2. Teachers and students should wear safety goggles whenever there is a possibility of flying objects or projectiles, such as when working with rubber bands.
3. Safety goggles used by more than one person should be sterilized between uses. One possible method of sterilization is to immerse the goggles in diluted laundry bleach followed by thorough rinsing and drying.
4. Proper precautions must be taken when using sharp objects such as knives, scalpels, compasses with sharp points, needles, and pins.
5. Students should not clean up broken glass. Teachers should use a broom and dustpan without touching the broken glass. Broken glass must be disposed of in a manner to prevent cuts or injury to the teacher, students, and custodial staff.
6. Teachers may decide to wear a laboratory apron or smock to prevent soiling or damage to clothing; if so, students should be similarly attired.
7. When working with hot materials, noxious plants, or live animals, teachers and students should wear appropriate hand protection.
8. Teachers and students should wash their hands upon completion of any experimental activity or at the end of the instructional session.

Safety with Fire and Heat Sources

1. Teachers should never leave the room while any flame is lighted or other heat source is in use.
2. Never heat flammable liquids. Heat only water or water solutions.
3. Use only glassware made from borosilicate glass (Kimax or Pyrex) for heating.
4. When working around a heat source, tie back long hair and secure loose clothing.
5. The area surrounding a heat source should be clean and have no combustible materials nearby.
6. When using a hot plate, locate it so that a child cannot pull it off the worktop or trip over the power cord.
7. Never leave the room while the hot plate is plugged in, whether or not it is in use; never allow students near an in-use hot plate if the teacher is not immediately beside the students.
8. Be certain that hot plates have been unplugged and are cool before handling. Check for residual heat by placing a few drops of water on the hot plate surface.

9. Never use alcohol burners.
10. Students should use candles only under the strict supervision of the teacher. Candles should be placed in a "drip pan" such as an aluminum pie plate large enough to contain the candle if it is knocked over.
11. The teacher should wear safety goggles and use heat-resistant mitts when working with hot materials. All students near hot liquids should wear safety goggles.
12. The teacher should keep a fire extinguisher near the activity area and be trained in its use.
13. The teacher should know what to do in case of fire. If a school policy does not exist, check with local fire officials for information.

Procedures for Using Dangerous Materials

1. Use only safety matches. Even safety matches should be used only with direct teacher supervision.
2. Use only non-mercury thermometers. Mercury from broken thermometers is difficult to clean up, and the vapors from spilled mercury are dangerous. Remember that thermometers are fragile; when students are handling them, supervise them so that the students won't use the thermometers as a stirring rod or allow them to roll off the table.
3. Store batteries with at least one terminal covered with tape.
4. Batteries exhibiting any corrosion should be discarded. Because the contents of batteries are potentially hazardous, batteries should not be cut open or taken apart. Check to see if batteries can be recycled in your area.
5. Never tell, encourage, or allow students to place any materials in or near their mouth, nose, or eyes.
6. Materials may include household chemicals. Before using household chemicals or other materials, study the label carefully to learn the hazards and precautions associated with such materials. Similarly, study the labels of chemicals purchased from a scientific supply house. The commercial suppliers of laboratory chemicals will furnish Material Safety Data Sheets (MSDSs) that describe the hazards and precautions for such materials in detail. These MSDSs should be on file in the school district office, and copies should be available in the classroom.
7. Do not touch "dry ice" (solid carbon dioxide) with the bare skin.
8. Always wear cotton or insulated gloves when handling dry ice. Do not store or place dry ice in a sealed container.
9. Liquid spills can be slippery. Clean up any spill immediately and properly as soon as it occurs. Follow the cleanup instructions given on the label or the MSDS for the substance.
10. Do not mix or use chemicals in any manner other than that stated in the approved procedure. At no time should a teacher undertake a new procedure without prior and full investigation of the chemical and physical properties of the materials to be used and of the outcomes of the proposed procedures. When planning to undertake a new procedure, it is a good practice to consult with a professional who is familiar with any potential problems.

Safety with Plants

1. Wash hands after working with seeds and plants. Many store-bought seeds have been coated with insecticides and/or fertilizers.

2. Never put seeds or plants in the mouth.
3. Do not handle seeds or plants if there are cuts or sores on the hands.
4. Some 700 species of plants are known to cause death or illness. Be aware of plants in the local area that are harmful. For more information, contact the local county agricultural agent.
5. Be aware of the signs of plant poisoning and act quickly if a student exhibits such signs after a lesson. Symptoms may include one or more of the following: headache, nausea, dizziness, vomiting, skin eruption, itching, or other skin irritation.
6. Be particularly alert to plant safety on field trips.

Safety with Animals

1. All handling of animals by students must be done voluntarily and only under immediate teacher supervision.
2. Students should not be allowed to mishandle or mistreat animals.
3. A safety lesson should be given to teach the students how to care for and treat the animal. A safety lesson on proper care and treatment of the animal should be given to students, ideally before the animal is brought into the classroom.
4. Animals caught in the wild should not be brought into the classroom.
5. For example, turtles are carriers of salmonella, and many wild animals are subject to rabies.
6. On field trips or during other outdoor activities, be aware of the danger of rabies exposure from wild animals. Also be aware of the potential hazards of insect bites, such as allergic reactions to bee stings or diseases spread by ticks or fleas.
7. At no time should dissection be done on an animal corpse unless it was specifically purchased for that purpose from a reliable supplier.
8. Any animal species that has been preserved in formaldehyde should not be used.

Emergency Procedures

1. Establish emergency procedures for at least the following: emergency first aid, electric shock, poisoning, burns, fire, evacuations, spills, and animal bites.
2. Evaluate each experimental procedure in advance of classroom use so that plans may be made in advance to handle possible emergencies.
3. Be sure that equipment and supplies needed for foreseen emergencies are available in or near the classroom.
4. Establish procedures for the notification of appropriate authorities and response agencies in the event of an emergency.

Disposal

Except for the disposal procedures described in the textbook in use, it is unlikely that any of the wastes generated in elementary science activities will be harmful to the environment. If the teacher has any questions concerning waste disposal, the science supervisor for the school or school district should be consulted.

Safety Awareness of Students

Safety instruction should begin at the earliest possible age. Students can begin to learn the importance of safety in the classroom, laboratory, and life in general at the elementary school level. The teacher must set the rules, but the teacher should also explain to the students why the rules are necessary. The students must also realize that anyone who does not follow the rules will lose the privilege of taking part in the fun, hands-on activities.

To reinforce the rules, teachers should engage the students in a discussion or activity. One activity could be a poster contest. The winning posters could be displayed in the room and used throughout the year to stress safety and enforce the safety rules.

General Safety Rules for Students

Science teachers always review the general safety rules with the students before beginning an activity.

1. Never do any experiment without the approval and direct supervision of teacher.
2. Always wear safety goggles when the teacher tells to do so. Never remove your goggles during an activity.
3. Know the location of all safety equipment in or near the classroom.
4. Never play with the safety equipment.
5. Tell the teacher immediately if an accident occurs.
6. Tell the teacher immediately if a spill occurs.
7. Tell the teacher immediately about any broken, chipped, or scratched glassware so that it may be properly cleaned up and disposed of.
8. Tie back long hair and secure loose clothing when working around flames.
9. If instructed to do so, wear the laboratory apron or smock to protect the clothing.
10. Never assume that anything that has been heated is cool. Hot glassware looks just like cool glassware.
11. Never taste anything during a laboratory activity. If an investigation involves tasting, it will be done in the cafeteria.
12. Clean up the work area upon completion of an activity.
13. Wash your hands with soap and water upon completion of an activity.

Helpful suggestions can be found in several resources. The teacher's edition of the textbook being used should have safety information on the activities. The state department of education should have publications available to assist with matters of safety and disposal. Many science supply houses offer safety and disposal publications. Expert advice can be obtained from organizations, such as the American Chemical Society, the Laboratory Safety Institute, the National Association of Biology Teachers, and the National Science Teachers Association. If a college or university is nearby, members of the science faculty are usually willing to assist in safety matters. (Addresses and telephone numbers of the organizations listed above appear at Appendix K.)

Science Activity 8-2 Develop a Safety Poster

Materials: Drawing materials (crayon and drawing paper)

Activity Procedure:
1. Divide the class into groups.

2. Ask them to develop a safety poster that will be on the classroom wall to remind the class of safety all through the semester.

3. Have students share their posters with the class.

Fundamentals for Effective Classroom Management

According to Kizlik (2018),

> classroom management and the management of student conduct are skills that teachers acquire and [struggle with] over time. Effective [science] teaching requires considerable skill in managing the myriad of tasks and situations that occur in the classroom each day. Skills such as effective classroom management are central to teaching and require "common sense," consistency, an often-undervalued teacher behavior, a sense of fairness, and courage. These skills also require that teachers understand, in more than one way, the psychological and developmental levels of their students. The skills associated with effective classroom management are only acquired with practice, feedback, and a willingness to learn from mistakes.

Fundamentals for effective classroom management from experienced classroom teachers who maintain an atmosphere that enhances learning are as follows (Kizlik, 2018):

- Know what you want and what you don't want
- Show and tell your students what you want
- When you get what you want, acknowledge (not praise) it
- When you get something else, act quickly and appropriately

In the constructivist science classrooms, management strategies are established as follows:

- Rules of behavior appropriate for expository settings should be enforced
- Teachers should move slowly from the expository mode to which children have become accustomed toward guided inquiry approaches.
- The teacher should demonstrate the activity before turning it over to the children. While doing so, the teacher should explain ways in which children can explore on their own. During this demonstration, the teacher should discuss behavior expectations and especially the safety precautions to be taken during the activity.
- While children are working on their own activities, a high degree of interaction exists between the teacher and individuals and small groups. Opportunities continually exist for the private encouragement of students to exhibit appropriate behavior and for the private correction of inappropriate behavior.

Through these management strategies, experienced science classroom teachers make opportunities to communicate with students and their parents as often as they can. By understanding each other better, it is easier to manage the classroom and the students. In Figure 8-2, the most important part is the finger touch that means the communication between the God and the Adam. To create your classroom environment well-managed and safe, science teachers are required to develop skills to communicate with their students and their parents.

FIGURE 8-2 Creation of Adam, by Michelangelo

Summary

Good science teachers are required to maintain and organize the classroom to make the learning environment safe. Maslow's Hierarchy of Needs is one of the strategies to see and understand students' actions, thoughts, and feelings that affect their lives. By satisfying students' physiological, safety, social, self-esteem, and actualization needs, the students in the classroom function properly, which brings more opportunities to communicate with the teachers. Also, family involvement in science classrooms is directly related to children's academic achievement and their attitudes toward science. The parents can employ their own culture to develop a friendly environment and have the children explore more. Therefore, this connection with the parents enables them to develop the respectful and comfortable learning environment where the children are familiar with but react with fear and apprehension when faced with the unfamiliar. Besides, the classroom environments that are oxygen-rich with animals and plants and science-centers facilitate successful science classrooms. Further, good science teachers need to develop skills to communicate with culturally and linguistically diverse students in science classrooms. The safety guidelines from ACS provide effective directions to make science classrooms safe.

Assignments

A musical is a total work of art. When a class creates a musical together and presents it to the public, like parents and teachers, the class will have more chances to communicate and understand each other. This assignment will make the class more organized and managed.

Steps:

1. In groups (same group with the lesson/unit),
   a. develop a story based on the topic of the unit;
   b. create a script (duration is about 10–15 minutes);
   c. add science songs/music;
   d. use technology for the stage background and background music/sound; and
   e. play it to the public (or record the musical and upload it on YouTube to share with the public)
2. Each group needs to submit a script, including
   a. the purposes of the musical;
   b. science concepts in the musical;
   c. casts' names;
   d. the plot of the story;
   e. script lines categorized by scenes; and
   f. the stage setting (including what kind of technology is used)

References

Alsubaie, M. A. (2015). Examples of Current Issues in the Multicultural Classroom. *Journal of Education and Practice,* 6(10). Retrieved April 13, 2018 from https://files.eric.ed.gov/fulltext/EJ1081654.pdf

American Chemical Society. (2014). "Safety in the Elementary (K–6) Science Classroom." Retrieved June 6, 2015 from https://www.acs.org/content/dam/acsorg/education/policies/safety/safety-in- the-elementary-k-6-science-classroom.pdf

Anderson, B. J., Eckburg, P. B., & Relucio K. I. (2002). *Alterations in the Thickness of Motor Cortical Subregions After Motor-Skill Learning and Exercise.* Cold Spring Harbor: Cold Spring Harbor Laboratory Press.

Bigelow, M. (2014). "Classroom Science Centers." NSTA Blog. Retrieved from http://nstacommunities.org/blog/2014/08/30/classroom-science-centers/

Brookover, W., Beady, C., Flood, P., Schweitzer, J., & Wisenbaker, J. (1979). *School Social Systems and Student Achievement: Schools can make a difference.* New York: Praeger.

Cotton, K., & Wikelund, K. R. 2001. "Parent involvement in education." *School Improvement Research Series.* Portland, OR: Northwest Regional Educational Laboratory. Retrieved June 6, 2015 from www.nwrel.org/scpd/sirs/3/cu6.html

Curtis-Fields, F. E. (2010). The Impact of Self-Efficacy, Locus of Control, and Perceived Parental Influence on the Academic Performance of Low and High Achieving African-American High School Children. DigitalCommons@Wayne State University. Retrieved from https://digitalcommons.wayne.edu/cgi/viewcontent.cgi?referer=https://www.google.com/&httpsredir=1&article=1064&context=oa_dissertations

Das, P. & Pattanaik, P. (2013). Self-esteem, locus of control and academic achievement among adolescents. *International Journal of Scientific Research in Recent Science.* 1(1), 1–5.

Desautels, L. (2014). *Addressing Our Needs: Maslow Comes to Life for Educators and Students.* Edutopia. Retrieved from: https://www.edutopia.org/blog/addressing-our-needs-maslow-hierarchy-lori-desautels

Ediger, M. (2012). "Recent Leaders in American Education," *College Student Journal,* 46(1), 174–177.

Hagedorn, M., O'Donnell, K., Smith, S., & Mulligan, G. (2008). *National Household Education Surveys Program of 2007: Data File User's Manual, Volume III, Parent and Family Involvement.* NCES Publication No. 2009-024. Washington, DC: U.S. Department of Education, National Center for Education Statistics.

Henderson, A. T., and K. L. Mapp. 2002. *A new wave of evidence: The impact of school, family, and community connections on student achievement.* Austin, TX: Southwest Educational Development Laboratory. Retrieved June 6, 2015 from www.sedl.org/connections/resources/evidence.pdf; conclusion available at www.sedl.org/connections/resources/conclusion-final-points.pdf.

Key, S. G. (2010). Diversity in Science Education. *Research into Practice,* Pearson Scott Foresman, 1–4.

Kim, J. (2012). *Defining and assessing parent empowerment and its relationship to academic achievement using the national household education survey: A focus on marginalized parents.* Unpublished doctoral dissertation. University of Maryland, College Park: MD.

Kim, E. (2002). The relationship between parental involvement and children's educational achievement in the Korean immigrant family. *Journal of Comparative Family Studies,* 33, 529–563.

Kim, K., & Rohner, R. P. (2002). Parental warmth, control, and involvement in schooling: Predicting academic achievement among Korean American adolescents. *Journal of Cross-Cultural Psychology,* 33, 127–140.

Kizlik, B. (2018) "Classroom Management Fundamentals." Adprima. Retrieved from http://www.adprima.com/managing.htm

Lauby, S. (2005). "Motivating Employees." American Society for Training and Development. Infoline, 2.

Madu, V. N. (2018). Locus of control, depressive symptoms, and perceived academic achievement of learners: A systemic review. *Global Journal of Educational Research,* 17(1). 25–56.

Ms. Mentor. (2009). "Living things in the Classroom," *NSTA Blog.* Retrieved June 6, 2015 from http://nstacommunities.org/blog/2009/10/08/living-things-in-the- classroom/

National PTA. (2014). "National Standards for Parent/Family Involvement Programs." Retrieved June 6, 2015 from http://www.doe.in.gov/sites/default/files/outreach/national-standards-parent.pdf

National Science Teacher Association (2009a). "Position Statement: Parental Involvement in Science Learning," Retrieved June 6, 2015 from http://www.nsta.org/about/positions/parents.aspx

National Science Teacher Association (2009b) "Living things in the classroom." NSTA Blog. Retrieved from: http://nstacommunities.org/blog/2009/10/08/living-things-in-the-classroom/

Parent Teacher Association (PTA). (1999). Position statement. Parent/family involvement: Effective parent involvement programs to foster student success.

Pate, P. E., and Andrews, P. G. (2006). Research summary: Parent involvement. Westerville, OH: National Middle School Association (NMSA). Retrieved June 6, 2015 from http://www.nmsa.org/Research/ResearchSummaries/ParentInvolvement/tabid/274/Default.aspx

Reynolds, A. & Shlafer, R. (2010). Recent Involvement in Early Education. Handbook of School–Family Partnerships. Routledge: Taylor & Francis Group. 158-174.

Welsh, J. (2013). "16-year-old Florida Honor Student Charged with Two Felonies for Doing a Science Experiment." *Business Insider.* Retrieved June 6, 2015 from http://www.businessinsider.com/kiera-wilmot-arrested-for-science-explosion- 2013-5

Tip of the Day 8: Why don't you add more to your classroom!

Hi future science teachers,

I found an interesting book, Sandy Bothmer's *Creating the Peaceable Classroom*. This book can give unique ideas about how to make your classroom peaceful to some of you who are struggling with classroom management. According to the book, first, you need to check if your classroom has the following:

1. Classroom pets
2. Objects to create movement (like a wind chime or a small fan behind a green plant that moves the leaves of the plant)
3. A science center with triangular signs
4. Sofas
5. Square throw pillows
6. Rugs
7. Round tables
8. Metal lamps
9. A fish tank
10. A fountain
11. A wheeled cart
12. Fabric with a wavy pattern
13. Wooden picture frames
14. A tree branch
15. Plants

How many of these do you have in your classroom? If your classroom is short of them, try to have all of the above in your classroom and then see if your classroom becomes peaceful. :-)

All these objects are Feng Shui elements (Fire, Earth, Metal, Water, and Wood). Feng Shui is an ancient Chinese practice derived from the awareness that the five elements of the natural world affect our being. Adding the Feng Shui elements to your classroom makes students feel connected to the natural world, thereby creating the peaceful classroom! Also, this is another cultural component to make your science classroom more culturally diverse.

Jiyoon

P.S. Explanations about each of the Feng Shui elements:

- *Fire*: The symbols of the element "Fire" include the color red, electricity, movement, and triangular shapes.
- *Earth*: Symbols of the element "earth" include the colors yellow, terra cotta, and brown or other earth tones; squares and cubes; feelings of safety, security, and peace; and the heart.
- *Metal*: Symbols of the element "Metal" include the colors of white, steel gray, metallic or reflective hues such as copper, gold, and silver, round shapes.
- *Water*: Symbols of the element "water" include the colors of black and other dark colors, glass, and wavy shapes
- *Wood*: Symbols of the element "Wood" include color green, tall and rectangular objects, wooden objects, and the experiences of growth

There are also many other cultural ways to make your classroom peaceful: breath, meditation, yoga, Reiki, and movement practices.

APPENDIX A

Professional Societies for Science Teachers

- American Association for the Advancement of Science (AAAS)
- American Association of Physics Teachers (AAPT)
- Association for the Education of Teachers in Science (AETS)
- Association for Science Education (ASE)
- International Society for Technology in Education (ISTE)
- National Academy of Sciences (NAS)
- National Science Teachers Association (NSTA)
- National Association for the Education of Young Children (NAEYC)
- National Association for Research in Science Teaching (NARST)
- National Association of Biology Teachers (NABT)
- National Association of Geoscience Teachers (NAGST)
- National Middle Level Science Teachers Association (NMSTA)
- Science Education Organizations on the Internet (SEOI)

APPENDIX B

Sites for Developing Questions

- http://planet.tvi.cc.nm.us/idc/webresources/Questioningtechniques.htm
  This site provides you with questioning techniques based on topics.
- http://www.stedwards.edu/cte/resources/blooms.htm
  This site emphasizes using questions to enhance learning. It provides you links related to questioning, Bloom's taxonomy, and questioning techniques.
- http://www.albanyacademyforgirls.org/Academics/RKWeb/questioning.htm
  This site provides you types of questions based on Bloom's taxonomy.
- http://www.justreadnow.com/strategies/bloom.htm
  This site gives you steps to Bloom's taxonomy questioning.
- http://www.unc.edu/learnnc/kinetic-connect/noframes.html
  This site gives you very good examples for questioning based on Bloom's taxonomy.
- http://questioning.org/articles.html
  This site gives you wonderful articles about questioning.
- http://honolulu.hawaii.edu/intranet/committees/FacDevCom/guidebk/teachtip/effquest.htm
  This site provides you with effective techniques for questioning.

APPENDIX C

5E Learning Cycle Model

- http://www.nationalcharterschools.org/uploads/pdf/resource_20040617125804_Using%20Inquiry.pdf
  This PDF file gives you details about each phase of the 5E instructional model, an example lesson, and resources.
- http://www.coe.ilstu.edu/scienceed/lorsbach/257lrcy.htm
  This site explains about the learning cycle and the 5Es.
- http://manzano.aps.edu/science/curriculum/planning.shtml
  This site gives you details about 5E instructional model: 5E lesson components, examples, and the learning cycle.

- http://www.personal.psu.edu/users/f/a/fah110/harrypotter.html
  This lesson gives you a very good example of the 5E instructional model.
- http://www.usd.edu/teachered/5-E%20lesson%20plan.doc
  This is another template for 5E lesson plan. With this template, you can develop a lesson thinking about what students do and what a teacher does.
- http://mdk12.org/instruction/curriculum/science/instruction.html
  This site provides two resources: explaining about the 5E model and designing a lesson using the 5E model.
- http://www.miamisci.org/ph/lpintro5e.html
  This site gives you an idea of the relationship between 5Es and constructivism.

APPENDIX D

Reading List Recommended by NSTA

- **Abayomi, the Brazilian Puma.** Darcy Pattison. Illustrated by Kitty Harvill. Mims House.
  A heart-warming story of an orphaned puma cub.
- **About Habitats: Forests.** Cathryn Sill. Illustrated by John Sill. Peachtree Publishers.
  Major types of forests depicted through simple text and full-page illustrations.
- **About Parrots. Cathryn Sill.** Illustrated by John Sill. Peachtree Publishers.
  Children can enjoy the beautiful, colorful illustrations while learning about these interesting birds found on five continents.
- **Amazing Giant Sea Creatures.** DK Publishing. DK Publishing.
  Enjoy an ocean adventure with giant sea creatures in this engaging fold-out flaps book.
- **Animalium. Jenny Broom.** Candlewick Press/Big Picture Press.
  Gallery-style illustrations of animals and habitats with descriptive supporting text.
- **Animals That Make Me Say OUCH!** Dawn Cusick. Charlesbridge/Imagine Publishing.
  A collection of animals with amazing defense adaptations that can inflict injury and even death if threatened.
- **Animals That Make Me Say WOW!** Dawn Cusick. Charlesbridge/Imagine Publishing.
  Our world is shared with animals that use amazing adaptations to help them survive with WOW skills.

- **At Home in Her Tomb.** Christine Liu-Perkins. Illustrated by Sarah S. Brannen. Charlesbridge.
  A true story of woman's body that was remarkably preserved for over 2000 years and was excavated to reveal a treasured time capsule.
- **A Baby Elephant in the Wild.** Caitlin O'Connell. Photos by Caitlin O'Connell and Timothy Rodwell. Houghton Mifflin Harcourt Books for Young Readers.
  Rare photos that depict the life of a baby elephant, weaving in facts, issues of survival, and social behavior.
- **Batman Science.** Tammy Enz and Agnieszka Biskup. Capstone/Capstone Young Readers.
  The Cape Crusader's as a scientist? The science behind the Dark Knight.
- **Beetle Busters.** Loree Griffin Burns. Photos by Ellen Harasimowicz. Houghton Mifflin Harcourt Books for Young Readers.
  A community's quest to combat an invasive beetle species in Massachusetts.
- **Behold the Beautiful Dung Beetle.** Cheryl Bardoe. Illustrated by Alan Marks. Charlesbridge.
  Incredible story of the importance of the dung beetles and their role on the earth.
- **Beneath the Sun.** Melissa Stewart. Illustrated by Constance R. Bergum. Peachtree Publishers.
  A wonderful look at how wild animals survive the heat in a variety of habitats.
- **Bone Collection: Skulls.** Camilla de la Bedoyere and Rob Colson. Illustrated by Sandra Doyle. Scholastic/Scholastic Paperback Nonfiction.

A fascinating collection of unique skulls featuring over 44 animals ranging from a humpback whales to human.

- **Chasing Cheetahs. Sy Montgomery.** Photos by Nic Bishop. Houghton Mifflin Harcourt Books for Young Readers.
Fascinating real life rescue in Namibia, Africa that saves lives every day.
- **Dinosaur!** DK Publishing. DK Publishing.
Always an exciting topic and this encyclopedia of dinosaurs is sure to engage students.
- **Drones.** Scholastic. Scholastic/Scholastic Paperback Nonfiction.
A multitude of facts and figures about 40 military and civilian drones.
- **Every Turtle Counts.** Sara Hoagland Hunter. Illustrated by Susan Spellman. Peter E. Randall Publisher.
A heart-warming story of an autistic girl who rescues a sea turtle and opens up a part of herself in the process.
- **Extreme Laboratories.** Ann O. Squire. Scholastic Library Publishing.
Labs and scientists in weird and wonderful places.
- **Eye to Eye.** Steve Jenkins. Illustrated by Steve Jenkins. Houghton Mifflin Harcourt Books for Young Readers.
A description of how the eyes of various animals allow them to survive.
- **Eyes Wide Open.** Paul Fleischman. Candlewick Press.
An aggressive and no holds bar book on the politics and science of climate change.
- **Feathers: Not Just for Flying.** Melissa Stewart. Illustrated by Sarah S. Brannen. Charlesbridge Publishing.
Feathers functioning as a forklift? Yes, and much more!
- **Full Speed Ahead!** Cruschiform. Abrams/Abrams Books for Young Readers.
Art and science meet in this beautiful book on the speeds of things.
- **Get the Scoop on Animal Puke.** Dawn Cusick. Charlesbridge/Imagine Publishing.
A great way to learn about animal anatomy through the study of regurgitation.

- **Handle with Care.** Loree Griffin Burns. Photos by Ellen Harasimowicz. Lerner Publishing Group/Millbrook Press.
Story of Costa Rican workers who care for the blue morpho butterfly.
- **Ivan.** Katherine Applegate. Illustrated by G. Brian Karas. Houghton Mifflin Harcourt Books for Young Readers.
A heart-wrenching story about a gorilla who has a very happy ending.
- **Neighborhood Sharks.** Katherine Roy. David Macaulay Studio/Roaring Brook Press.
Great whites come to the Farallon Islands.
- **Next Time You See a Maple Seed.** Emily Morgan. Illustrated by Steven David Johnson, Tom Uhlman, and Todd Amacker. NSTA Kids/an imprint of NSTA Press.
Beautiful close-up photography of the life cycle of the maple seed or samara.
- **Ocean.** DK Publishing. DK Publishing.
A breathtaking book about every aspect of the ocean.
- **Park Scientists.** Mary Kay Carson. Photos by Tom Uhlman. Houghton Mifflin Harcourt Books for Young Readers.
Explore our National Parks, America's natural laboratories and living museums, through the eyes of a field scientist.
- **Polar Bears and Penguins.** Katharine Hall. Arbordale Publishing.
Beautiful photographs of Polar Bears and Penguins and how they are similar and different.
- **Sally Ride.** Sue Macy. Simon & Schuster/Aladdin.
Inspiring Biography of this brave, dedicated astronaut.
- **Secrets of the Sky Caves.** Sandra K. Athans. Lerner Publishing Group/Millbrook Press.
Follow a mountaineer on this incredible journey discovering caves built into high steep cliffs filled with mummies and other artifacts.
- **Sniffer Dogs.** Nancy F. Castaldo. Houghton Mifflin Harcourt Books for Young Readers.
A description of how dogs (many of which have come from shelters) are being trained to help their human companions in searches, scientific research by tracking animals, and alerting for serious medical conditions.

- **Star Stuff.** Stephanie Roth Sisson. Macmillan Children's Publishing Group/Roaring Brook Press.
  A beautifully written and illustrated picture book introduces to a young Carl Sagan and his journey into the sciences.
- **Super Human Encyclopedia.** DK Publishing. DK Publishing.
  Reference book that includes photos, diagrams, and graphs on the human body.
- **Super Sniffers.** Dorothy Hinshaw Patent. Bloomsbury/Walker Children's Books.
  Exciting jobs that dogs have to help people keep the world safe.
- **The Griffin and the Dinosaur.** Marc Aronson and Adrienne Mayor. Illustrated by Chris Muller. National Geographic Society.
  The journey of Adrienne Mayor in explaining what mythological creatures may have actually been by researching, observing, and inferencing.
- **The Next Wave.** Elizabeth Rusch. Houghton Mifflin Harcourt Books for Young Readers.
  An examination of the various ways scientists are attempting to convert ocean wave energy into a renewable energy source.
- **The Planets.** DK Publishing. DK Publishing.
  Take a tour of the planets in this beautiful photographed and illustrated guide to our solar system.
- **Tiny Creatures.** Nicola Davies. Illustrated by Emily Sutton. Candlewick Press.
  Great elementary view into the world of microbes.
- **Tooling Around.** Ellen Jackson. Illustrated by Renné Benoit. Charlesbridge.
  Animal Engineers are described by mixing poetry with explanations.
- **Ultimate Bodypedia.** Patricia Daniels, Christina Wilsdon, and Jen Agresta. National Geographic Society.
  An amazing tour of the many complexities of the human body, inside and out!
- **Wild About Bears.** Jeannie Brett. Illustrated by Jeannie Brett. Charlesbridge.
- *Informative and beautifully illustrated, this book compares the eight species of bears found around the world.*

SOURCE

http://science.nsta.org/publications/ostb/ostb2015.aspx

APPENDIX E

Rubric for Assessing Interactive Qualities of Open Educational Resources

The rubric shown below in Table E-1 is modified from the model of Roblyer and Ekhaml (2000) and the Constructivism Approach (Wilson & Peterson, 2006) and has three separate dimensions that contribute to the level of interaction/interactivity and construction of knowledge.

RUBRIC DIRECTIONS: The rubric shown below has three separate elements that contribute to the level of interaction and interactivity. For each of these three elements, circle a description below it that applies best to the interactive OER that you find online. After reviewing all elements and circling the appropriate level, add up the points to determine the level of interactive qualities (e.g., low, moderate, or high). The site that has the highest score is the best site for your lesson.

- Low Interactive qualities 1–7 points
- Moderate interactive qualities 8–14 points
- High interactive qualities 15–20 points

TABLE E-1 Rubric for Assessing Interactive Qualities of Open Educational Resources.

Scale (point)	Element #1 Social Rapport-building Activities	Element #2 Instructional Designs for Learning	Element #3 Levels of Constructivism Approach
Few interactive qualities (1 point)	The **interactive** OERs do not encourage students to get to know one another on a personal basis. No activities require social interaction.	The **interactive** OERs do not require two-way interaction between instructor and students; they call for one-way delivery of information (e.g., instructor lectures, text delivery).	The **interactive** OERs do not encourage students to construct new knowledge based on their own individualized experience, Not emphasizing the application of knowledge in real life situations and caring multi-intelligence
Minimum interactive qualities (2 points each)	**The interactive** OERs provide for exchanging personal information among students.	The **interactive** OERs require students to communicate with the instructor on an individual basis only.	The **interactive** OERs encourage students to construct new knowledge based on their own individualized experience, but not emphasizing the application of knowledge in real life situations and caring multi-intelligence

Moderate interactive qualities (3 points each)	The **interactive** OERs provide more than one activity designed to increase social rapport among students.	In addition to the requiring students to communicate with the instructor, the **interactive** OERs require students to work with one another.	The **interactive** OERs enable students to construct new knowledge based on their own individualized experience, emphasizing the application of knowledge in real life situations but not caring multi-intelligence
Above average interactive qualities (4 points each)	The **interactive** OERs provide several activities designed to increase social rapport among students.	In addition to the requiring students to communicate with the instructor, the **interactive** OERs require students to work with one another (e. g., in pairs or small groups) and share results with one another and the rest of the class.	The **interactive** OERs enable students to construct new knowledge based on their own individualized experience, providing a few opportunities to apply their knowledge in real life situations and care multi-intelligence
High level of interactive qualities (5 points each)	In addition to providing for exchanges of personal information among students, **the interactive** OERs provide a variety of activities designed to increase social rapport among students.	In addition to the requiring students to communicate with the instructor, the **interactive** OERs require students to work with one another (e. g., in pairs or small groups) and outside experts and share results with one another and the rest of the class.	The **interactive** OERs enable students to construct new knowledge based on their own individualized experience, emphasizing the application of knowledge in real life situations and caring multi-intelligence
Total for each:	_____ pts.	_____ pts.	_____ pts.
Total overall:	_____ pts.		

TABLE E-2 **Interactive Open Educational Resources.**

Contents	Physical Science	Life Science	Earth/Space Science
interactive OERs	• 101 in 1 Physics Solver • Active Sonar • Alchemy Glossary • AP Physics • Atom in a Box • Chemical Equation • Chemical Formulas • Chemistry Formulas • Chemistry Terms • Colour Collider • Dictionary of Chemistry • Dictionary of Physics • Elemental Table Formulary: Physics • Gear Ratio	• 3D Brain • 3D Cell Simulator • 3D4Medical • A Life Cycle App • Anatomy and Physiology • Anatomy Flash • Bio Dictionary • Biology Memory • Biology Body Parts - Human Body • Reproduction • Bugs and Insects • Bugism • Buzz Aldrin Portal to Science	• 8 Planets • Astronomy HD • Beautiful Planet • Cosmic Discoveries • Deep Sky • Earth Observer • EnchantedLearning/Astronomy • ExoplanetGravity Balls • Google Earth • Grand Tour 3D: Pocket Solar System • HD Astronomy • HD Solar System • Highered.mheducation • Jupiter Study Guide • Mars Globe HD • Mars Study Guide

• iChemistryLab • iLab Timer • Molecules • Mr Science Show • Mythbusters • Newton's Cradle Physics • Newton's Laws • Oxford Dictionary of Chemistry • Pendulums • Periodic • Periodic Table • Physics Formulas • Physics Puzzles • Physics XL • Physiology Glossary • Wolfram General Chemistry Course • XChem • Touch Physics • Toy Physics • Titration Simulator • Rocket Universe • Glencoe/Science Virtual Labs • Vital Lab/Ohio University: Chemical and Physical change lab • Quia/pop	• Cellular Biology Digestion • Frog dissection • Genetic Decoder • Genetic History • Genetics Study Guide • HD Marine Life • Human Anatomy • iAnatomy • Insects HD • Marine Life • Nanosaur 2 • Nature Human Genome • Rat Dissection • Respiratory System • Virtual Frog Dissection • ZeroBio • Livebinders	• Moon Globe HD • National Geography: Reason for Seasons • Orion StarSeek PRO • Planets • Planet's New • Pluto Study Guide • Saturn Study Guide • Scienceu/observatory • Solar System • Solar System Guide • Space Images • SpaceTime for iPad • Star Chart • Star Gazer • StarMap 3D Plus • Star New • Stars • Star Walk • Stellarium XL • SchoolMedia interactive • The Weather Channel • Solar System Simulation • Bootslearningstore: Sunshine and Shadow • Science Games for Kids: Sun, Light, & Shadows • E-learning for Kids: Science – Scotland-Sun & Shadows • Why Do We Have Seasons Interactive • UNL Astronomy: Sun's Rays Simulator • UNL Astronomy: Seasons and Ecliptic Simulator • StudyJams/Science: Rocks & Minerals • NASA: Earth Observatory • Venus Study Guide • WeatherBug • Weather Channels • Weatherwizkids

SOURCE

Yoon, J. (2017). Developing a List and a Rubric of Interactive Open Education Resources (OER) For Science Teacher Candidates of Diverse Students. *TEM Jounral, 6*(3). 512–524.

APPENDIX F

Instructions for Developing Lesson Plans for Korean Students

This task is to develop a lesson plan that delivers *multicultural/diversity objectives* while employing an *inclusive teaching strategy or strategies for* a Korean Science classroom. Your task is to:

- Create a lesson plan using the 5E learning cycle
- Develop and add multicultural/diverse objectives into your existing simple science lesson plan
- Insert one or more activities to learn the multicultural/diverse objectives
- Respond to the following and attach answers to your lesson plan:
  - What makes this lesson for a multicultural/diverse classroom?
  - What services did or would you provide for non-English speaking students in this classroom?

Multicultural/Diverse Lesson Plan Rubric

FACTOR 1: LESSON PLAN OBJECTIVES

This factor evaluates a teacher candidate's ability to design a meaningful multicultural/diversity objectives lesson. Low scores usually indicate that the student omitted a multicultural/diversity objective. Often the multicultural/diversity objectives component appears to be an add-on to an existing plan just to meet the requirement. Higher scores were awarded if multicultural/diversity objectives were more central to the lesson, engaged attitudes and beliefs, and went beyond factual information. At the proficient level, the lesson plan synthesizes personal reflection, addresses developmental adaptations, and includes skills relevant outside the classroom.

FACTOR 2: LESSON PLAN MECHANICS

This factor evaluates the ability to synthesize lesson plan objectives with classroom activities and assessments. Low scores represent weak or non-existent links between these components. This usually occurs because student teachers omitted assessment, assessed informally, evaluated by single means, or with a select group of their students. High scores represent stronger, multiple, and inclusive activities and assessments.

FACTOR 3: LESSON PLAN RATIONALE

This factor evaluates the ability to competently explain why his or her lesson plan was multicultural/diversity objectives and inclusive. Low scores indicate the misconception that a multicultural/diversity objectives lesson employs culturally stereotyped activities. While "developing" scores acknowledge diversity, "proficient" scores include societal perspectives.

FACTOR 4: LESSON PLAN INCLUSIVENESS

This factor evaluates the ability to provide inclusive teaching strategies, especially to individuals whose primary language was not English. Low scores convey no adaptations or the single approach. Lesson plans with a "developing" score reflect more responsibility for addressing student needs, typically with varying instructional strategies. A "proficient" lesson plan promotes the classroom as a community of learners with multiple learning styles.

The following rubric is modified from the four-factor rubric of Ambrosio (2002) for evaluating teachers in developing science lessons for Korean students.

SOURCE

Ambrosio, A.L., Sequin , C.A. , & Hogan, E.L. (2002). "Assessing performance-based outcomes of multicultural/diversity objectives lesson plans: A component within a comprehensive teacher education assessment design." Multicultural/Diversity Objectives Perspectives, 3(1), 15–22.

APPENDIX G

Teacher License Preparation Materials

- Full Option Science System (FOSS) Curriculum Matrix: http://lawrencehallofscience.org/foss/scope/index.html

  The FOSS curriculum matrix requires that teacher candidates can summarize what the science contents are for K–8 graders.
- FOSS Teacher Preparation Video: http://lhsfoss.org/fossweb/schools/teachervideos/index.html

  FOSS has developed videos to assist teachers in preparing science lessons. They include teacher guide information, safety precautions, set-up procedures, and classroom footage. Transcripts for each video are also available.
- Educational Testing Service (ETS) Materials: http://cms.texes-ets.org/prepmaterials/

  A wide variety of test preparation resources are available on the site. Teacher candidates can download Preparation Manuals for TExES (Texas Examinations of Educator Standards), TExMaT (Texas Examinations for Master Teachers), and TASC/TASC-ASL (Texas Assessment of Sign Communication-American Sign Language) tests, as well as other publications.
- Preparation of Elementary School Teachers to Teach Science in California: http://www.ccst.us/publications/2010/2010K-6.pdf

  The site provides information about how to prepare elementary school teachers to teach science in California. Even though the procedure will be different depending on states, it gives the major ideas of the preparation of elementary science education.
- Texas Education Agency (TEA) Teacher Preparation Site: http://cms.texes-ets.org/texes/prepmaterials/texes- preparation-manuals/

  The test preparation manuals are designed to help teacher candidates prepare for the TExES test for teacher certification. This site is to familiarize the teacher candidates with the competencies to be tested, test question formats, and pertinent study resources.

APPENDIX H

Organizations for Safety Advice

American Chemical Society, Chemical Health and Safety Referral Service
> 1155 Sixteenth St., NW
> Washington, DC 20036
> (800) 227-5558, ext. 4513
> LIBRARY@acs.org
> chemistry.org

American Chemical Society Committee on Chemical Safety
> 1155 Sixteenth St., NW
> Washington, DC 20036
> chemistry.org/committees/ccs

National Association of Biology Teachers
> 12030 Sunrise Valley Dr., Suite 110
> Reston, VA 20191
> (703) 264-9696
> www.NABT.org

National Science Teachers Association
> 1840 Wilson Blvd.
> Arlington, VA 22201-3000
> (703) 243-7100
> www.NSTA.org

Laboratory Safety Institute
> 192 Worcester Rd.
> Natick, MA 01760
> (508) 647-1900
> www.labsafety.org

Image Credits

Design 0-1: Source: https://commons.wikimedia. org/wiki/File:Creaci%C3%B3n_de_Ad%C3%A1n_ (Miguel_%C3%81ngel).jpg.

Figure 4-2: Adapted from: "Science in the Community," Helping Your Child Learn Science, pp. 36-42. U.S. Department of Education, 2005. Copyright in the Public Domain.

Figure 6-4: Mechanics Magazine, 1824.

Figure 6-7: William Ely Hill, Puck, v. 78, no. 2018, pp. 11.

Figure 7-2: David Jerner Martin, "Figure 8.4," Elementary Science Methods: A Constructivist Approach, pp. 279. Copyright © 2011 by Cengage Learning, Inc.

Figure 7-4: Elizabeth A. Desy, Scott A. Peterson, and Vicky Brockman, Table 1 from: "Gender Differences in Science-Related Attitudes and Interests Among Middle School and High School Students," Science Educator, vol. 20, no. 2, pp. 24. Copyright © 2011 by National Science Education Leadership Association. Reprinted with permission. Provided by ProQuest LLC. All rights reserved.

Table 8-2: Arthur J. Reynolds and Rebecca J. Shlafer, "Table 7.1," Handbook of School-Family Partnerships, ed. Sandra L. Christenson and Amy L. Reschly, pp. 160. Copyright © 2010 by Taylor & Francis Group. Reprinted with permission.

Figure 8-2: Source: http://commons.wikimedia.org/ wiki/File:Creation_of_Adam_Michelangelo.jpg.

www.ingramcontent.com/pod-product-compliance
Lightning Source LLC
Chambersburg PA
CBHW081549220326

41598CB00036B/6621